Chromatographic Separation
and Extraction with
Foamed Plastics and Rubbers

CHROMATOGRAPHIC SCIENCE

A Series of Monographs

Editor: JACK CAZES
Fairfield, Connecticut

Volume 1: Dynamics of Chromatography (out of print)
J. Calvin Giddings

Volume 2: Gas Chromatographic Analysis of Drugs and Pesticides
Benjamin J. Gudzinowicz

Volume 3: Principles of Adsorption Chromatography: The Separation of Nonionic
Organic Compounds (out of print)
Lloyd R. Snyder

Volume 4: Multicomponent Chromatography: Theory of Interference (out of print)
Friedrich Helfferich and Gerhard Klein

Volume 5: Quantitative Analysis by Gas Chromatography
Joseph Novák

Volume 6: High-Speed Liquid Chromatography
Peter M. Rajcsanyi and Elisabeth Rajcsanyi

Volume 7: Fundamentals of Integrated CG-MS (in three parts)
Benjamin J. Gudzinowicz, Michael J. Gudzinowicz, and Horace F. Martin

Volume 8: Liquid Chromtography of Polymers and Related Materials
Jack Cazes

Volume 9: GLC and HPLC Determination of Therapeutic Agents (in three parts)
Part 1 edited by Kiyoshi Tsuji and Walter Morozowich
Parts 2 and 3 edited by Kiyoshi Tsuji

Volume 10: Biological/Biomedical Applications of Liquid Chromatography
Edited by Gerald L. Hawk

Volume 11: Chromatography in Petroleum Analysis
Edited by Klaus H. Altgelt and T. H. Gouw

Volume 12: Biological/Biomedical Applications of Liquid Chromatography II
Edited by Gerald L. Hawk

Volume 13: Liquid Chromatography of Polymers and Related Materials II
Edited by Jack Cazes and Xavier Delamare

Volume 14: Introduction to Analytical Gas Chromatography: History, Principles, and
Practice
John A. Perry

Volume 15: Applications of Glass Capillary Gas Chromatography
Edited by Walter G. Jennings

Volume 16: Steroid Analysis by HPLC: Recent Applications
Edited by Marie P. Kautsky

Volume 17: Thin-Layer Chromatography: Techniques and Applications (in press)
Bernard Fried and Joseph Sherma

Volume 18: Biological/Biomedical Applications of Liquid Chromatography III
Edited by Gerald L. Hawk

Volume 19: Liquid Chromatography of Polymers and Related Materials III
Edited by Jack Cazes

Volume 20: Biological/Biomedical Applications of Liquid Chromatography IV
Edited by Gerald L. Hawk

Volume 21: Chromatographic Separation and Extraction with Foamed Plastics and Rubbers
G. J. Moody and J. D. R. Thomas

Other Volumes in Preparation

Chromatographic Separation and Extraction with Foamed Plastics and Rubbers

G. J. Moody

J. D. R. Thomas

Chemistry Department
University of Wales Institute of Science and Technology
Cardiff, Wales

MARCEL DEKKER, INC.   New York and Basel

Library of Congress Cataloging in Publication Data

Moody, G. J.
    Chromatographic separation and extraction with
foamed plastics and rubbers.

    (Chromatographic science ; v. 21)
    Bibliography: p.
    Includes index.
    1. Chromatographic analysis. 2. Plastic foams.
3. Foam rubber. I. Thomas, John David Ronald.
II. Title. III. Series. [DNLM: 1. Chromatography.
2. Polyurethanes. Wl CH943 v.21 / TP 1180.P8 M817c]
QD79.C4M66 1982   543'.0892   82-4636
ISBN 0-8247-1549-7

MARCEL DEKKER, INC.
270 Madison Avenue, New York, New York 10016

Current printing (last digit):
10 9 8 7 6 5 4 3 2 1

PRINTED IN THE UNITED STATES OF AMERICA

Preface

Polyurethanes are manufactured annually in multimillion kilogram quantities for a wide range of markets. Such availability
merits their exploitation for separation and concentration
processes in chemistry.

The first chapter in this book deals with the broad principles of some typical preparations of the polyurethane foams
and their important physical and chemical properties. Chapters
2, 3, and 4 are devoted exclusively to their applications in
analytical chemistry. In 1970, Dr. H. J. M. Bowen of Reading
University, U.K., reported the use of polyurethane foam for the
facile extraction of a diverse range of inorganic and organic
substances. The considerable research effort subsequently
expended on their application in the preconcentration of such
materials--often from samples containing extremely low levels--
is covered in Chapters 2 and 3. In particular, the assay of
air- and water-borne pesticides has been greatly simplified
following their preconcentration on foams. For this purpose,
the polyester or polyether foams may be employed as synthesized,
or after modification by physically coating with liquid ion-
exchangers or organophosphorus liquid extractants, while
chemical grafting is also practiced.

Finally, their use in gas-solid chromatography, gas-liquid chromatography, and low-pressure liquid-liquid chromatography is described in Chapter 4. In this regard the resolution of cis- and trans-isomers of metal chelates by gas-chromatography is particularly noteworthy.

We wish to thank the publishers for their cooperation and patience and Mrs. P. Bevan for her excellent secretarial services.

G. J. Moody
J. D. R. Thomas

Contents

Preface   iii

1 Polyurethanes: Nature and Properties   1

   1.1 Introduction   1
   1.2 The Fundamental Chemistry of Polyurethane
        Synthesis   2
   1.3 Types of Polyurethanes   4
   1.4 The Synthesis of Polyurethanes   6
   1.5 The Physical and Chemical Properties of
        Polyurethane Foams   12
   1.6 Health Hazards Associated
        with Polyurethane Foams   14

2 Inorganic Applications of Polyurethane Foams
   in Aqueous Media   19

   2.1 Unloaded Foams   19
   2.2 Loaded Foams   49
   2.3 Silicone Rubber-Loaded Foams   78

3 The Concentration of Organic Compounds from
   Dilute Samples by Polyurethane Foams   79

   3.1 Pesticides   79
   3.2 Herbicides   102
   3.3 Phthalate Esters   104
   3.4 Alkyl Benzene Sulfonate   106

3.5   Phenols   107
3.6   Nicotine   110
3.7   Oils   111
3.8   Enzyme Inhibitors and Immunoadsorption
      of Cells   113

4   Foam Chromatography of Organic Compounds   115

4.1   Gas-Solid Chromatography   115
4.2   Gas-Liquid Chromatography   119
4.3   Liquid-Liquid Chromatography   120
4.4   Modified Polyurethane Foams   123

References   129

Index   135

Chromatographic Separation and Extraction with Foamed Plastics and Rubbers

1

Polyurethanes:
Nature and Properties

1.1 INTRODUCTION

Gas chromatography is a widely practiced technique which has
facilitated an astonishing range of successful resolutions of
complex mixtures. Numerous materials are available as column
supports for this purpose. One of them, namely, foamed polyurethane, was first disclosed in 1967 by Van Venrooy [1] in a
U.S. patent and facilitated the resolution of a mixture comprising heptane, benzene, toluene, ethylbenzene, pentyl benzene and
2-ethyl phenol in 22 min with helium carrier gas. Since then
the same support has been used for a wide variety of separations
[2-11], for example, aliphatic hydrocarbons, chloroanilines,
phenols, aliphatic alcohols and halides, and transition metal
complexes.

In 1970 Bowen reported that such polyurethane foams were
also capable of absorbing a variety of metal ions from dilute
aqueous solutions [12]. Subsequently, their versatility for
the concentration and separation of materials, especially metal
ions [13-60] and organopesticides [49,58-81], has been firmly
established. Polyurethane foams have also found limited use in
studies of the stabilization of cholinesterase [82-84], the

immunoadsorption of cells [85], the removal of phenols in solutions [86], nicotine in cigarette smoke [87], and oil from
spillages [88,89].

1.2 THE FUNDAMENTAL CHEMISTRY OF POLYURETHANE SYNTHESIS

Polyurethanes are substituted amide esters of carbamic acid,
$RN-\overset{H}{\underset{|}{C}}\overset{O}{\overset{\|}{C}}-OH$, and are synthesized by the condensation-polymerization
of polyesters, or polyols, with diisocyanates and can be repre-
sented by the general equation

$$OCNRNCO + HOR'OH \rightarrow OCNRN\overset{H}{\underset{|}{C}}\overset{O}{\overset{\|}{C}}OR'OH \tag{1}$$

Here, R is commonly a low-molecular-mass alkyl or aryl moiety
and R' is typically an alkyl polyester or polyether chain. This
basic urethane unit $-\overset{H}{\underset{|}{N}}-\overset{O}{\overset{\|}{C}}-O-$ can then react further to give a
long, linear polymer. This chain propogation is achieved by
further similar condensations:

$$OCNRN\overset{H}{\underset{|}{C}}\overset{O}{\overset{\|}{C}}OR'OH + OCNRNCO \rightarrow OCNRN\overset{H}{\underset{|}{C}}\overset{O}{\overset{\|}{C}}OR'O\overset{O}{\overset{\|}{C}}\overset{H}{\underset{|}{N}}RNCO \tag{2}$$

or with a variety of agencies, for example, water or diamines.
Thus, reaction with water forms an unstable carbamic acid,

$$RNCO + H_2O \rightarrow R-\overset{H}{\underset{|}{N}}-\overset{O}{\overset{\|}{C}}-OH \tag{3}$$

which decomposes to an amine and carbon dioxide:

$$R-\overset{H}{\underset{|}{N}}-\overset{O}{\overset{\|}{C}}-OH \rightarrow RNH_2 + CO_2 \tag{4}$$

This isocyanate-water interaction is exploited for foam formation since the carbon dioxide liberated functions as a convenient
in situ blowing agency. Silicones are also added as foam stabi-
lizers.

The amine, from reaction (4), can also interact further
with isocyanate entities and results in chain extension with
the insertion of urea bridges:

$$RNH_2 + OCNRNCO \rightarrow R-\overset{\overset{H}{|}}{N}-\overset{\overset{O}{\parallel}}{C}-\overset{\overset{H}{|}}{N}-R-NCO \qquad (5)$$

On the other hand, the carbamic acid may react with
isocyanate,

$$R-\overset{\overset{H}{|}}{N}-\overset{\overset{O}{\parallel}}{C}-OH + OCNRNCO \rightarrow R-\overset{\overset{H}{|}}{N}-\overset{\overset{O}{\parallel}}{C}-O-\overset{\overset{O}{\parallel}}{C}-\overset{\overset{H}{|}}{N}-R-NCO \qquad (6)$$

and the anhydride from Eq. (6) decomposes to give further
blowing agency and urea bridges:

$$R-\overset{\overset{H}{|}}{N}-\overset{\overset{O}{\parallel}}{C}-O-\overset{\overset{O}{\parallel}}{C}-\overset{\overset{H}{|}}{N}-R-NCO \rightarrow R-\overset{\overset{H}{|}}{N}-\overset{\overset{O}{\parallel}}{C}-\overset{\overset{H}{|}}{N}-R-NCO + CO_2 \qquad (7)$$

However, in addition to the primary chain propagation steps,
side reactions lead to branching and cross-linking:

$$(8)$$

The synthesis is thus complex with many possible reaction
pathways resulting in the formation of products with biuret and
allophanate units:

$$R-\overset{\overset{H}{|}}{N}-\overset{\overset{O}{\parallel}}{C}-\overset{\overset{H}{|}}{N}-R + R-NCO \rightarrow R\overset{\overset{H}{|}}{\underset{}{N}}-\overset{\overset{O}{\parallel}}{C}-\overset{\overset{R}{|}}{N}-\overset{\overset{O}{\parallel}}{C}-N-H\overset{}{}R \qquad (9)$$

Biuret

$$R-\overset{\overset{H}{|}}{N}-\overset{\overset{O}{\parallel}}{C}-OR' + RNCO \rightarrow R\overset{\overset{H}{|}}{\underset{}{N}}-\overset{\overset{O}{\parallel}}{C}-\overset{\overset{R}{|}}{N}-\overset{\overset{O}{\parallel}}{C}-O\overset{}{}R' \qquad (10)$$

Allophanate

1.3   TYPES OF POLYURETHANES

Depending on the nature of the diisocyanate-polyol condensation
Eq. (1), the resulting polymer can take one of several forms,
namely, flexible foams, rigid foams, synthetic rubbers, coatings,
adhesives, fibers, paints, and molding compounds. All these
products come under the general heading of polyurethanes. Some
are highly elastic with outstanding abrasion resistance as well
as high resistance to tearing and extension--all features which
are extremely important in their analytical applications.

     Polyurethanes have been categorized in a variety of ways.
On the basis of cell structure, urethanes are either open-cell
foams possessing intricate interconnected cellular structures,
through which liquids and gases can pass, or closed-cell foams
with separate nonconnecting gas cells. On the other hand, foams
may be described as either high-density or low-density materials
depending on the relative proportion of gas cells to solid polyurethane. Foams with densities <0.48 g cm^{-3} are described as
low-density foams.

     Yet another classification is described in terms of flexible
or rigid foams wherein the former types exhibit resilience and
flexibility when subject to deformation, whereas the latter type
is more resistant to deformation stress. Flexible and rigid
foams tend to have open- and closed-cell structures, respectively
[90]. The closed-cell structures in rigid foams generally comprise interconnected dodecahedra with up to 90% of the closed
cells having intact walls or membranes. The usual open-cell
(i.e., flexible) foam comprises an array of blown pentagonal
dodecahedral cells based on interconnecting struts (or strands).
At least two windows (or membranes) in each cell must be rup-
tured if fluids are to pass freely through the foam matrix [4,
18].

Reticulated polyurethane foam produced by hydrolysis of the membranes in the open-cell foams comprise windowless cells containing only strands [91].

Thus, cell structure relates to the presence or absence of windows, i.e., the number of windows per cell depends on the experimental conditions at the time of condensation.

The structure of open-pore polyurethanes consist of agglomerated spherical particles (5 to 10 μm diameter) rather than the above-mentioned interconnected struts left from blown dodecahedral cells in the foamed urethanes. The pore sizes, generally <10 μm, are considerably smaller than the finest foam cells known. Yet the spherical particles are bonded to each other to give a rigid, highly permeable matrix which facilitates high solvent flow rates (Fig. 1).

Such urethanes can be further modified, e.g., by coating with stationary phases such as Apiezon and Carbowax [6], liquid ion-exchangers or mixing with charcoal [72,73,75]. Techniques have also been developed for the production of ion-exchanger foams [92] and redox foams [23].

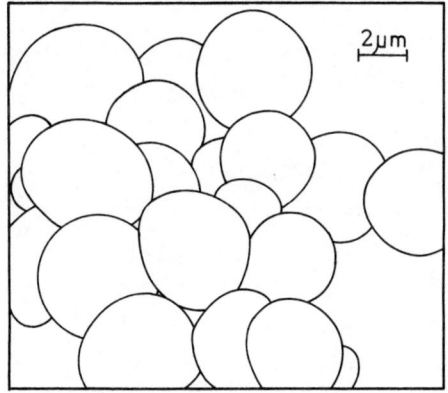

FIGURE 1 Reproduction of photomicrograph of a typical open-pore polyurethane foam at $\sim$ ×5500.

Most of the published work on the analytical application
of cellular plastics has been based on such products. Foamed
polyurethanes have found extensive use in the area of collection,
recovery, and separation of organic and inorganic materials from
aqueous media, while open-pore polyurethane has been largely
employed for column chromatography. These applications will be
covered in Chaps. 2 through 4, while the remainder of this
chapter is devoted to more specific details of foam preparation,
their chemical and physical properties, and some potential health
hazards.

1.4 THE SYNTHESIS OF POLYURETHANES

1.4.1 The Basic Ingredients

The product prepared from the oxypropylation of diethylenetri-
amine and commercially available from Union Carbide Corporation
as LA-475 is widely used as a starting polyol and whose tertiary
nitrogens provide a self-contained catalyst for the condensation
reaction with the polyisocyanate [6].

$$CH_3\overset{\overset{\displaystyle OH}{|}}{C}CH_2CH_3$$
$$N[CH_2CH_2N(CH_2\overset{\overset{\displaystyle OH}{|}}{C}HCH_3)_2]_2$$

Such polyols should have a hydroxyl functionality of at
least four [2], i.e., the average number of OH groups per
molecule, which for the above formula is five.

Other polyols which can be employed include polyethylene
glycols, 1,4-butanediol, trimethylol phenol, glucose, caprolac-
tone, and Quadrol, a Wyandotte Chemical Corporation product
based on the oxypropylation of ethylenediamine.

Polyhydroxyl-terminated polyesters may also be employed
but are generally more viscous and costly than the polyether-
type polyols.

One widely used isocyanate, and designated MDI, is based
on a mixture of polyarylpolyalkylene polyisocyanates is available
from Mobray Chemical Corporation. This product, Mondur MR, comprises 4,4'-diphenylmethane diisocyanate (40 to 50%) with lesser
quantities of ter-, tetra-, and pentaisocyanates, respectively:

Typically, the values of n for crude batches of MDI range from
0.5 to 2 and hence for n = 0.5 results in a functionality of
2.5. The highest functionality F is often located in the polyol,
but polyisocyanates with F > 2 can be realized in different MDI
batches by raising the quantities of higher molecular mass components to give F $\sim$ 3.2, for example [2].

Many isocyanates can be employed in conjunction with a
suitable polyol but preferably should be liquid at the temperature (usually ambient) of the condensation reactant. The most
widely used reagent is toluene diisocyanate (TDI) and is sold as
a 80:20, or 65:35, mixture of 2,4- and 2,6-toluenediisocyanate.
Other viable isocyanates include 4,4'-diphenylmethane diisocyanate
(MDI), 4-4'-diphenylisocyanate (XDI), 1,5-naphthalene diisocyanate
(NDI), and hexamethylene diisocyanate (HMDI).

These diisocyanates must be protected from moisture attack
owing to the formation of amines, Eq. (4). The quality of the
prospective diisocyanate can be checked by infrared spectroscopy,
and any batch showing broad bands at 3326 cm^{-1} (-NH stretch)
should be discarded [93].

Thus, a range of both polyol and polyisocyanates has been
used, and depending on the type of polyurethane intended, one

or more other ingredients may also be included, e.g., plasticizer, inhibitor, chemical blowing agent, stabilizer, and
catalyst. For example, the preparation of foams requires
some agency for blowing or expanding the product. One well-
established technique is based on the carbon dioxide formed by
the hydrolysis of isocyanate with water added at the time of
the reaction, Eq. (4). Alternatively, nitrogen or air may be
whipped into the polymerizing mass [87] or a fluorinated hydro-
carbon added prior to the reaction, alone (e.g., 25 parts of
trichlorofluoromethane to 100 parts of polyol) or with a little
water [52].

Again, where necessary, a wide variety of catalysts has
been employed, e.g., trimethylamine, N,N'-diethylpiperazine,
tribenzylamine, and tin(II) compounds such as tin octoate,
dibutyltin dioctoate, and dibutyltin acetate. The amount of
catalysts is far less than that used for blowing agents,
typically 0.02 to 2 parts per 100 parts of polyol, unless the
amine catalyst is to be chemically incorporated into the polyurethane product. One technique [87] is to dissolve the catalyst in a Freon, which is then added to the polyol, followed
by isocyanate, when foaming ensues.

1.4.2 The Synthesis of Open-Pore Polyurethanes (OPP)

The formation of OPP material depends on the relatively slow
precipitation of a polyurethane foam from a quiescent homogeneous
diluted mixture of chosen reactants [4]. In particular, reaction
times are influenced by temperature, catalyst, solvent, reactant
concentration, and the NCO/OH ratio. These parameters can be
adjusted to assess the most favorable gelation period. The term
gelation describes the change of state from the initial, clear
RNCO/ROH solution to a gel. This phenomenon is often apparent
as an opacity or may be detected by the onset of a rapid rise
in viscosity [3]. Many different reaction conditions can be

chosen for the synthesis of OPP foams (or indeed the other foam types already described). Each one will yield products with different physical properties, e.g., density, surface area, hardness, particle size, porosity, and compression strength.

Jefferson and Salyer [2] have described some typical preparations of open-pore polyurethane foams which amply illustrate the effect of varying certain parameters on the physical properties of final products, e.g., NCO/OH ratios, concentration of polyol and isocyanate, and nature of solvents (Tables 1 and 2).

Thus, the polyol LA-475 (100 g) with F $\sim$ 5 and crude isocyanate MDI (116 g) with F $\sim$ 2.5 were each separately dissolved in about equal masses of toluene having the total masses shown for runs 1 to 7 in Table 1 and corresponding to 12 to 25% of the reactants in toluene. The respective toluene solutions were mixed, briefly stirred (<1 min) until homogeneous, poured into a mold, and left undisturbed while polymerization continued. After $\sim$4 hr the cured, toluene-loaded product was removed from the mold and the toluene allowed to evaporate [2]. Virtually identical procedures were followed in other runs with carbon tetrachloride and mixed solvents, but in run 14 xylene was finally removed by vacuum distillation [2].

The porosities of the OPP foams are controlled by the concentration of isocyanate + polyol for the seven toluene runs and to a lesser degree for the four carbon tetrachloride runs, while in both sets of runs the densities increased as the concentration of solids increased from 12 to 25%. The influence of solvent(s) at 25% solids (runs 7, 12, and 14) is reflected in the density and shore A hardness values, rather than porosity of product (Table 1).

The effect of the NCO/OH ratio on the final foam material for two different starting concentrations of solids is shown in Table 2.

TABLE 1 Some Properties of Polyurethane Foams Prepared from the Polyol LA-475 and Crude MDI

Run	LA-475 + MDI (mass %)[a]	(g)	Solvent (g)	Property of Polyurethane Foam					
				Density (g cm^{-3})	Surface area (m^2 g^{-1})	Shore A hardness[b]	Particle size (μm)	Porosity[c] (%)	Compression strength[d] (psi)
1	12	1584	Toluene	0.14		0		90	6
2	15	1224		0.18	0.5	0		86	
3	17	1054		0.24		0	1.6	82	370–385
4	17.8	1000		0.24		10		81	
5	18.5	952		0.24		10		81	
6	20	864		0.27	0.8	28		81	
7	25	649		0.38		88		75	300
8	12	1584	CCl$_4$	0.26		0		79	
9	15	1224		0.32		40		77	
10	17	1054		0.33		57		74	
11	20	864		0.38		65		74	
12	25	600	Toluene + CCl$_4$ (1:1 m/m)	0.36		79		74	
13	17.7	200	Toluene + JP-4 (1:1 m/m)	0.17		0	6.2	89	
14	25	600	Xylene	0.24		0		80	

[a]Each of the reaction mixtures (1 to 14) corresponds to an OH/NCO ratio of unity.
[b]Measured by ASTM D676.
[c]Measured using a Beckman Model No. 930 Air Comparison Pycnometer.
[d]Measured at 10% deflection by ASTM D1621-59.
Source: Ref. 2.

TABLE 2 Some Properties of Polyurethane Foams Prepared from
the Polyol LA-475 and Crude MDI

		Property of Polyurethane Foams		
Solids (mass %)[a]	NCO/OH	Density (g cm^{-3})	Shore A hardness[b]	Porosity (%)
18	1.05	0.23	--	86
	1.00	0.24	10	81
25	1.00	0.38	88	75
	0.97	0.38	90	74
	0.95	0.33	85	77
	0.90	0.41	91	70

[a]Dissolved as separate solutions in toluene (cf. Table 1).
[b]Measured by ASTM D676.
Source: Ref. 2.

The NCO/OH functionality ratio serves to indicate the
stoichiometry of the condensation reaction and is also related
to the index format which for the reaction conditions in Table 2
ranges from 90 to 105.

Isocyanate functionality appeared to exert little influence
on the density or porosity of OPP product (Table 3).

TABLE 3 Properties of Polyurethane Foams Prepared from Crude
MDI with Three Different Functionalities

		Property of Polyurethane Foam	
Solids (mass %)[a]	Crude (F)	Density (g cm^{-3})	Porosity (%)
15	2.3	0.17	89
	2.5	0.18	86
	3.2	0.18	88

[a]Dissolved as separate solutions in toluene at NCO/OH ratio 1.0.
Source: Ref. 2.

1.5   THE PHYSICAL AND CHEMICAL PROPERTIES OF POLYURETHANE FOAMS

Both the physical and chemical properties of foams relate to
the method of syntheses and by the structure and molecular mass
of polyol, the diisocyanate, and the degree of branching or
cross-linking. The choice of polyol exerts a major influence
on the final foam product, especially the flexibility or rigid-
ity [94]. Thus, flexible foams (low cross-link density) arise
from polyols of moderately high molecular mass with low degree
of branching, whereas rigid foams (high cross-link density) can
be made from polyols of low molecular mass and with more branching.

The resistance to solvent attack increases in polyurethanes
with high cross-linked density and appears to be unaffected by
the type of aromatic diisocyanate and is reduced by using a large
excess of diisocyanate [95]. The extent of the allophanate network increases with reaction time and temperature for TDI-based
urethanes. These allophanate linkages exert a major influence
on the mechanical strength of foams.

The relative proportion of allophanate as well as urea,
urethane, and biuret linkages, the amount of free, i.e., unreacted, NCO, and the identity of foams of unknown compositions
have been studied using infrared spectroscopy [96-100] and
nuclear magnetic resonance (nmr) techniques [101,102]. Thus,
the infrared bands in the 1600 to 1800 cm^{-1} range permit the
assay of allophanate concentration, while the strong band at
~ 2270 cm^{-1} is assigned [97] to unreacted, i.e., free, isocyanate
groups. In general, TDI-based foams contained less free NCO than
their MDI analogs [98,99]. For a sorbitol polyol/TDI-based foam
prepared with indexes from 95 to 110, the NCO content fell more
rapidly over a 4-week postpreparative period, possibly due to
the greater reactivity of TDI [98]. Even after this 4-week
period, the 95 index foam (i.e., underindexed) still contained

∿0.2 mass % free NCO groups, which in turn indicates the presence of unreacted OH-sorbitol groups [98]. Such residual excess NCO units can be effectively hydrolyzed with hot water or steam [87], but otherwise will remain to react with atmospheric water, especially in humid climates.

Sumi and co-workers reported that urethane, allophanate, and biuret linkages in polyurethane foams could be discriminated from each other on the basis of chemical shifts measured in dimethyl sulfoxide or N,N-dimethylacetamide [101]. The nmr spectra of several TDI-based foams have been run in arsenic trichloride between 100 and 110°C when the signals in the 7.0- to 8.0-ppm region indicated that the 2,4- and 2,6-isomeric ratio are essentially the same as that of the starting TDI mixture [102]. Good quality nmr spectra were also obtained using trifluoroacetic acid instead of arsenic trichloride in which some foam degradation was evident during runs over 2 to 3 days [102].

The physical resilience of polyurethanes is well matched by their inertness toward a wide range of chemicals. Bowen [12,13] reported that six different CO_2-blown foams [based on TDI(80:20)/polypropane-1,2-diol, and with a twofold density and fourfold surface area range] are dissolved by concentrated sulfuric acid, oxidized by alkaline permanganate and destroyed by concentrated nitric acid and bromine vapor. However, apart from reversible swelling, the six foams studied remained mostly unaffected by water, light petroleum, benzene, carbon tetrachloride, chloroform, diethyl ether, diisopropylether, acetone, 4-methylpentan-2-one, ethylacetate, isopentyl acetate, alcohols, hydrochloric acid (<6 M), sulfuric acid (<4 M), nitric acid (<2 M), ammonia (2 M), sodium hydroxide (2 M), and acetic acid (2 M). Open-pore polyurethanes are also compatible with organic solvents and water and exhibit some weak-base anion-exchange capacity [9]. An open-pore polyurethane product based on MDI and a polyether (molecular mass 4800), however, did not disintegrate, at least

within 24 hr, in concentrated hydrochloric acid, sulfuric acid
(9 M), or 88% phosphoric acid [52].

Bowen found that the foams degraded when heated to between
180 and 220°C and also slowly turned brown in ultraviolet light
[12,13]. Thermogravimetric analysis conducted on open-pore
foams (ρ = 0.154 g cm^{-3}) revealed the onset of decomposition
at 200°C, although discoloration was noticeable below this
temperature [4].

Generally foams prepared from aromatic isocyanates tend
to yellow on prolonged exposure to sunlight which may promote
autoxidation of the polymers to give imide/quinone-type materials. On the other hand, the aliphatic-based analogs are less
prone to photochemical effects and have better thermal stability.

The most efficient foam for the recovery of insecticides
and polychlorinated biphenyls from waters has been judged on
the maximum uptake of methylene blue per kilogram of foam during
24 hr [62]. This simple spectroscopic technique has been utilized
[60] to follow the exposure of a foam to a variety of conditions
and is recommended as a simple check on the quality of a foam
batch which is to be used over a long period. Thus a fresh
batch of polyurethane foam D2 [based on a trifunctional ethylene
oxide/propylene oxide copolymer and TDI (80:20) blown with carbon
dioxide and trichlorofluoromethane] absorbed more methylene blue
than the same foam exposed to an Hanovia Model 16 ultraviolet
lamp for 48 hr at a distance of 40 cm, to direct sunlight for
35 days, or after intermittent use over several years (Fig. 2).
The uptake of transition metal thiocyanate complexes from aqueous
solutions at pH 5.5 followed the same trend [60].

1.6 HEALTH HAZARDS ASSOCIATED WITH POLYURETHANE FOAMS

Inhalation of isocyanate as vapor or fine particles may induce
human respiratory problems which can be temporary, or worse,
induce a sensitivized condition requiring that exposed persons

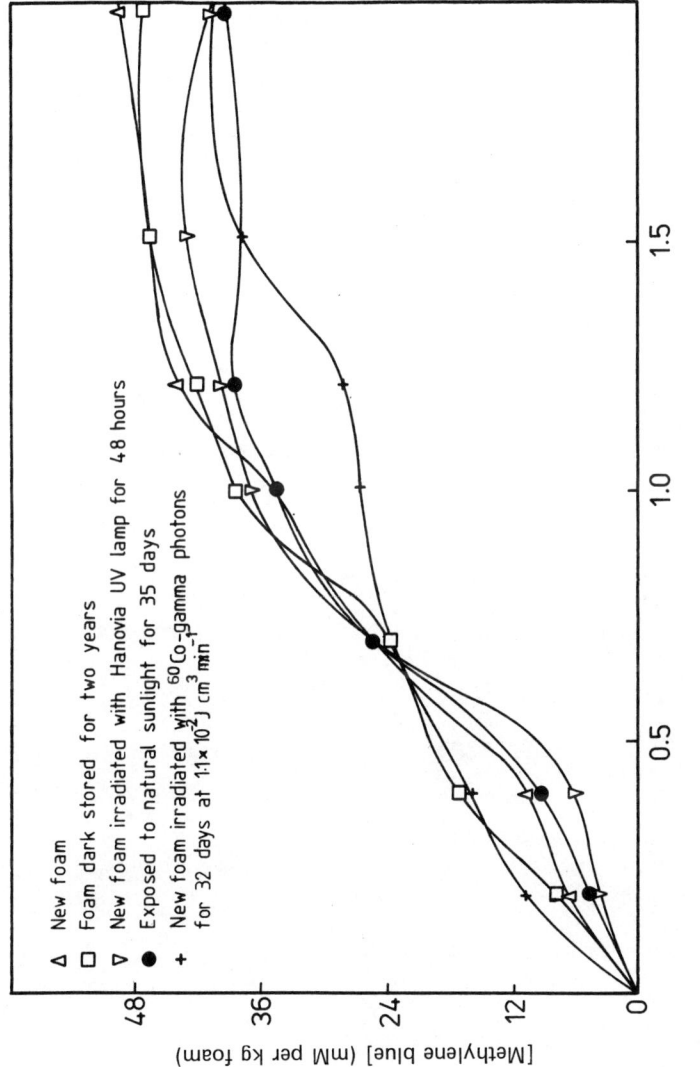

FIGURE 2 The uptake of methylene blue by D2 polyether foam after various treatments.

△ New foam
□ Foam dark stored for two years
▽ New foam irradiated with Hanovia UV lamp for 48 hours
● Exposed to natural sunlight for 35 days
+ New foam irradiated with ^{60}Co–gamma photons for 32 days at $1.1 \times 10^2 \text{ J cm}^3 \text{ min}^{-1}$

[Methylene blue] (mM)

[Methylene blue] (mM per kg foam)

be transferred to a "non-isocyanate" area. These hazards relate
not only to monomer isocyanates such as the ubiquitous TDI (whose
threshold limit was lowered to 0.02 ppm in 1961) but possibly to
any molecular species containing free NCO groups.

Exposure to TDI causes irritation as well as immunological
sensitization of the respiratory tract using animal models.
Recovery after a 3-hr exposure to TDI was extremely slow, and
cumulative effects could be observed at levels >0.02 ppm [103].
Similar, reproducible patterns were evident following 3-hr expo-
sures to monoisocyanate analogs of TDI and HDI, while aromatic
and aliphatic diisocyanates had comparable potency [104].

Polyurethanes are now manufactured in considerable quantities for a wide range of products including carpets, upholstery,
insulation, adhesives, paints, and printing inks. As a result,
a large number of industrial personnel are potentially at risk,
and, of course, the same is true for those engaged on analytical
applications, except that the foam is frequently purchased on a
ready-to-use basis. The hazards and the techniques for analysis
of low levels of isocyanates have recently been reviewed [105,
106].

The presence of 2,4-diaminotoluene in flexible TDI-based
polyurethanes is also of concern since it is included in the
NIOSH list of potential carcinogens. In addition, the amine
content of foam is a factor in their color stability. Guthrie
and McKinney [107] have developed a sensitive method for the
detection of low levels of the 2,4- or 2,6-isomers in foams
which arise by hydrolysis of the -NCO groups:

(11)

TABLE 4 Diaminotoluene (DAT) Content of Seven Flexible
TDI-Based Polyurethane Foams

	DAT (ppm)	
Foam sample	2,4-Isomer	2,6-Isomer
Hydrophilic polyether, type 1	20	0
Hydrophilic polyether, type 2	77	0
Hydrophilic polyether, type 3	6	0
Hydrophobic polyether, type 1	67	80
Hydrophobic polyether, type 2	338	0
Hydrophobic polyether, type 3	99	0
Hydrophobic polyether, type 4	123	62

Source: Ref. 107. Reprinted with permission from J. L. Guthrie
and R. W. McKinney, *Analytical Chemistry*, 49:1676 (1977).
Copyright 1977, American Chemical Society.

Each of the seven foams contained the 2,4-isomer, but only
two contained 2,6-diaminotoluene (Table 4). Moreover, the DAT
content depends on the sample zone, being more near the outside
of a large foam than inside. The method based on rapid methanol
extraction of foams, separation of DATs by TLC, and in situ
fluorimetric assay is also applicable to amines in urethane
products from isocyanates other than TDI [107]. Recently,
Edwards [108] has determined the free MDI in polyurethane prepolymers using HPLC followed by infrared detection at 2270 cm^{-1}.
Free toluene diisocyanate, which elutes after the MDI peak, can
also be determined in the prepolymers, but no separation of the
2,4- and 2,6-TDI isomers was possible.

2

Inorganic Applications of Polyurethane Foams in Aqueous Media

In 1964 Lal et al. [109] described the extraction of trace elements from seawater using iron(III) hydroxide supported on natural sponge materials. Aluminium, beryllium, silicon, and titanium could thus be extracted by towing sponges impregnated with iron(III) hydroxide through coastal waters [109]. Bowen in 1970 was the first to observe that a number of materials could be extracted from aqueous media by synthetic polyurethane foams loaded with diethyl ether [12]. Since then, foamed polyurethanes have been utilized in numerous, diverse analytical applications mostly for the collection, separation, and recovery of inorganic, as well as organic, materials from aqueous systems. This chapter is concerned with the inorganic applications of unloaded foams and foams modified by loading with liquid ion-exchangers, liquid extractants, chelating agents, and inorganic precipitates and by radiation grafting.

2.1 UNLOADED FOAMS

Bowen's classic research [12] was conducted with 27 different foams, each being a copolymer of toluene diisocyanate and poly-propane-1,2-diol, and with variable bulk densities and surface

areas, e.g., 0.0154 to 0.0297 g cm^{-3} and 7.6 to 32.5 $m^2 kg^{-1}$,
respectively. The extractions were effected with simple
apparatus, wherein typically a known mass of foam M_F in the form
of small cubes ($\sim$1-cm sides) was shaken with a known mass of
solution M_S ($M_S \sim 10^2 M_F$) for about 3 hr in a stoppered flask
to establish solute-foam equilibrium. Instead of shaking, the
foam cubes could be alternatively compressed and released (about
10 times) with a glass rod to ensure maximum solute contact,
while for large-scale recovery a solution (or indeed vapor)
could be repeatedly passed through a foam compressed as a
cylinder in a confined glass tube.

The uptake of material by the foam sample was generally
quantified indirectly by analysis of the solution at equilibrium
using radiotracer, spectrophotometric, and titration techniques
(Table 1). Surface areas of the initial foams were also measured by a shaking equilibrium technique using 1-[^{14}C]stearic
acid in n-heptane solvent which is itself, like n-hexane,
probably absorbed to a negligible extent [110]. There was no
correlation between surface area and bulk density for sixteen
of the foams investigated [110].

The distribution ratio D for 15 different substances
(Table 1) was calculated from the percentage of solute extracted
(E) by the foam using the expression

$$D = \frac{M_S}{M_F} (100/E - 1)$$

Neither D nor capacity values were constant but varied with
the relative masses of polyurethane/aqueous phase, and in most
cases the absorption isotherms were not linear since the foam
became rapidly saturated with solute and the most favorable
absorption profiles were obtained with more dilute solutions
due to saturation of absorption sites on the foam. Because of
the large numbers of foams and solutes chosen for investigation,

TABLE 1 Absorption Capacities and Distribution Ratios for 27 Different Polyurethane Foams

Compound	Solvent	Capacity (mol kg^{-1})	Distribution ratio (D)	Nature of species absorbed on foam	Technique for assay of residual aqueous species at equilibrium
Mercury(II)	0.2 M HCl	$\geq$ 0.06	30–125	b	0.28 MeV-γ photon/^{203}Hg
Iron(III)	6 M HCl	0.38–0.77	40–1400	$FeCl_n^{-}$ (n = 3.85)	1.10 MeV-γ photon/^{59}Fe or absorbance of $FeSCN^{2+}$ at 425 nm
Rhenium(III)		$\geq$ 0.43	340[a]	b	Absorbance at 520 nm
Thallium(III)	0.2 M HCl	0.29–0.46	170–3700	$TlCl^{-}$ (n = 4.02)	Bremmstrahlung (<0.15 MeV) ^{204}Tl
Gold(III)	0.2 M HCl	0.33–1.27	45–3900	$AuCl_n^{-}$ (n = 4.35)	⎫
Gold(III)	0.2 M HBr	0.71–1.07	140–570[a]	$AuBr_n^{-}$ (n = 4.20)	0.41 MeV-γ photon/^{198}Au or absorbance at 290 nm
Gold(III)	0.2 M HI	0.76–1.68	250–590[a]	AuI_n^{-} (n = 3.82)	⎭
Antimony(V)	6 M HCl	0.25–0.49	75–500	$SbCl_n^{-}$ (n = 2.63)	0.60 MeV-γ photon/^{124}Sb
Molybdenum(VI)		0.10–0.42	27–41[a]	b	0.74 MeV-γ photon/^{99}Mo
Uranium(VI)	Al(NO$_3$)$_3$ (saturated)	$\geq$ 0.16	25–100	b	β or γ counts
Iodine	Water	0.67–1.36	700–3500	I_2	Thiosulphate titration
Chloroform		0.17–0.25	250–420	$CHCl_3$	Pyridine complex at 560 nm
Benzene		0.45–1.79	60–140	C_6H_6	Absorbance at 254 nm
Phenol		$\geq$ 0.032	45–410	C_6H_5OH	Absorbance at 270 nm
Styrene		--	2000[a]	b	Absorbance at 262 nm

[a] Measurements based on just one foam sample.
[b] Not studied.
Source: Refs. 12 and 13.

Bowen found it impracticable to measure the entire absorption
isotherm for every foam sample, and moreover, different foam
samples presented different isotherms, which relates to the
range of distribution ratios and capacity values in Table 1 for
any single chemical species.

The uptake of iodine and gold(III) was higher at 22°C than
at 51 and 44°C, respectively, whereas the reverse was evident
for iron(III) at 51°C compared with runs at 22°C. Antimony(V)
was absorbed to the greatest extent on a foam with a surface
area of 20.8 m^2kg^{-1} compared with a foam of 32.5 m^2kg^{-1}, but in
both instances exceeded the uptake by a foam of 20.3 m^2kg^{-1}.
On the other hand, the uptake of thallium(III) decreased as the
surface area of the foams increased [12].

Mercury could be reextracted with dilute alkali, iron(III)
with water, or dilute ammonia, but iron(III) thiocyanate was
strongly absorbed, gold(III) partially with ethyl acetate,
uranium(VI) with water, and iodine with acetone or carbon
tetrachloride [13].

In addition, Bowen established that the following substances
were strongly absorbed by the foams: chlorobenzene, bromobenzene,
iodobenzene, carbon tetrachloride, iron(III) thiocyanate, lead
dithizonate and mercury diphenylcarbazone from water, copper
dithizonate from hydrochloric acid (0.1 M), and zinc dithizonate
from ammonium acetate buffer at pH 8.5. Negligible or poor
absorption was evident for acetone, methyl iodide, n-hexane,
sulfur dioxide, ammonium phosphate, copper(I) and (II) acetate,
nickel nitrate, palladium chloride, silver nitrate, and tetra-
phenylarsonium permanganate [12].

At 70°C foam absorbed about its own mass of iodine from the
vapor state, some 90% of which could be recovered by passing
steam through the foam [13]. At room temperature, foams were
effectively saturated after absorbing about 10% benzene, 50%
chloroform, or 30% trichloroethylene by mass, and their recoveries were managed by heating to ∿70°C in an air stream [13].

The surface areas of foams imply that 1 kg of a foam sample should absorb 0.1 to 0.3 mmol of solute compared with typical values of 1.5 mol kg^{-1} (Table 1). Hence, the mechanism must involve absorption, not adsorption processes. This was confirmed by microscopic examination of foam samples exposed to both aqueous iodine or iodine vapor when the iodine color was uniformly distributed throughout the foam section and not just confined to the surface [12].

The nature of the species on some of the foams was investigated by diffuse reflectance spectroscopy [12]. The spectrum of iodine-loaded foam shows a broad band between 397 and 562 nm and is quite different from the profile of aqueous iodine which shows a maximum at 345 nm with a shoulder between 420 and 500 nm. The spectra of absorbed gold(III) and iron(III) halides qualitatively resembled the spectra of analogous complexes in acid solution and which are believed to be tetrachloro anions. Analysis of certain complexes on the dried foams, after absorption of gold or iron, lends further support to the spectroscopic evidence. The stoichiometries listed in Table 1 were determined by boiling the complexes with alkali and titrating the liberated bromide or chloride with mercury(II) nitrate and diphenylcarbazone indicator. The nature of the absorbed species for antimony(V) was considered to be a mixed hydroxy or oxochloro anion.

Bowen listed two classes of solutes which are strongly absorbed on the various foams, namely, those which exist in aqueous solution as free molecules with high polarizability, e.g., metal dithizonates and iodine and the polarizable univalent anions such as tetrachloro iron(III) and tetrachloro gold(III) species [12]. The mechanism of the absorption could involve ion exchange or chelation via oxygen sites in the polymer network. However, Bowen stressed a significant point, namely, that these tetrachloro anions are themselves also readily extracted into diethyl ether from aqueous acid media. The foam

being a polyether should similarly take up protons followed by
anions to maintain charge balance [12].

Bowen's fundamental work [12,13] formed the basis for the
subsequent research effort on the application of polyurethane
foams to inorganic systems, especially at Eötvös University,
Hungary, and Manitoba University, Canada. In this post-1970
connection, the following points are especially noteworthy
regarding Bowen's research:

1. Simple apparatus was used: the alternating compression-
    release cycle of foam immersed in the aqueous sample with
    glass rods was really the forerunner of the automated
    devices developed at Eötvös and Manitoba universities.
    Techniques for stripping absorbed material from foams were
    also described, and suggestions were proposed for the
    recovery of metal complexes from very dilute industrial
    waste.
         Their use to extract various substances from an air
    stream was also outlined (see Chap. 3).
2. Preliminary work with alternative polyurethane foams
    incorporating chelating agencies, e.g., dimethylglyoxime
    and phenylthiourea was described.
3. Mechanistic aspects of complex-foam interactions were pro-
    posed and the stoichiometries of the species absorbed on
    the foam analyzed.
4. Techniques for estimating the extent to which an aqueous
    species is extracted by foams were devised, especially
    using spectrophotometry and radiotracers.

A selection of studies on inorganic systems is presented
in Table 2 for unloaded foams, although some work reported for
foams loaded with various components, e.g., tributyl phosphate,
is supplemented with comparative results on the analogous
unloaded foams [16,20,36,38,39,42,52,54]. As a general rule,
foams are subjected to some pretreatment washing scheme and then
equilibrated with the sample species by a batch or column method.
At equilibrium it has been common practice to estimate the uptake
of material by the foam on the basis of the material remaining
in the solution.

The direct determination of absorbed species, e.g., iodine,
has also been reported using ^{131}I as radiotracer [34,41,43,44].

Other absorbed species have been similarly determined using ^{198}Au [53], ^{124}Sb [111] and ^{119m}Sn [57]. Tanaka et al. estimated the concentration of an absorbed alkylbenzene sulfonate/
crystal violet complex by direct visual comparison with a series
of standards [112], while foams loaded with various organic
reagents have facilitated the direct detection of metal ions by
the color formed [26]. More recently x-ray fluorescence has
been employed in the same direct mode for the quantitative assay
of the Co(NCS)$_4^{2-}$ complex absorbed on polyether foams [58].
Copper, iron, lead, nickel, and zinc did not interfere even at
the lowest level (0.05 ppm Co) studied since their complexes
are either unstable or are not actually absorbed on the foam as
in the case of lead and nickel [58].

Each approach is based on the analysis at the equilibrium
point, as assessed from previous experience or on a miss-hit
basis for a new system. Ion-selective electrodes, on the other
hand, permit a continuous in situ assessment of the depletion of
a solution containing foam and have been used to follow the
uptake of barium and calcium by organophosphorus-loaded polyether foams [60].

2.1.1   Precious Metals [12,13,16,20,21,46,50,53]

Bowen proposed that polyurethane foam might be a valuable
vehicle for the recovery of rare or precious metals from very
dilute solutions [13]. Subsequently, he investigated the use
of such foams for the recovery of gold from the barren waste
($\sim 1.25 \times 10^8$ kg per year) of the Tjikotok mineral processing
plant in West Java, containing gold (trace), silver (0.8 mg),
and sodium cyanide (800 mg) per kilogram of waste [50]. Silver
recoveries were not reported, but about 50% of the gold was
recovered using a batch technique.

Schiller and Cook found that 99.4% of the gold in a solution could be removed by shaking with Dunlopillo foam (649/33
polyether type) for 90 min compared with a sorption efficiency

TABLE 2   Selected Studies of Metal Ion/Polyurethane Foam Systems

Source and type of foam	Pretreatment and shape/size of foam
Acid resistant type A or C (Union Carbide)	Soaked in 1 M HCl for 24 hr with intermittent squeezing. Washed with water. Extracted with acetone in Soxhlet for 6 to 12 hr.
Regular foam (Local Stores)	Air or vacuum dried. Stored in dark.
Polyester DiSPo (Winnipeg)	Same procedure except 0.1 M HNO$_3$ used in some cases instead of 0.1 M HCl.
Polyether (Dunlop)	Washed with 1 M HCl, water, acetone, and dried. Small pieces (∿3 mm). Unspecified.
Polyurethanes (unspecified)	Cubes (∿2-mm edge). Washed with 1 M HCl water, acetone, and air-dried
Open-cell polyethers H1 and H2 (N. Hungarian works, Sajobabony)	
Open-cell polyethers W8100, 8300, and 8600 (Kunstoffbüro GmbH, München)	Cut into small pieces. Soaked in 2 M HCl. Washed in water. Washed in acetone.
Open-cell polyether, Greiner KG, Schaumstoffwerk-Kremsmünster (Austria)	Oven dried at 80°C.
Open-cell polyester PPI-800nFR Eurofoam, Damstraat (Belgium)	Same procedure except 0.2 M HCl used to prevent ester hydrolysis.

[a] Atomic absorption spectroscopy.
[b] Ion-selective electrodes.

Typical equilibrium technique	System	Analysis technique	Ref.
Cylinders (e.g., 4 cm in diameter and 1.5 cm long). Automatically squeezed and released below level of sample solution at 22 ± 3°C; also some column runs.	$IrCl_6{}^{2-}$ } $PtCl_6{}^{2-}$ }	Spectrophotometry and ^{192}Ir	[46]
	$SbCl_3$ } $SbCl_5$ }	^{124}Sb	[111]
	Co(II)/ thiocyanate {	XRF ^{60}Co AA^a	[58] [116]
	Sn(II); Sn(IV)	^{119m}Sn	[57]
	Ga(III) }		[48]
Plugs (3.2 × 3.2 cm) in glass columns	Ga(III) } Fe(III) }	AA^a	[47,115, 114]
Batch	Group II metals and transition metals	AA^a, ISE^b and spectrophotometry	[60]
Column	$AuCl_3$ }	^{198}Au	[53] [50]
Batch	Au-thiourea complex }		[16] [20]
	In(III)	^{114m}In	[42]
Batch and column investigations for the seven metal ions listed opposite and in aqueous thiocyanate solutions	Hg(II)	^{203}Hg	[42]
	Zn(II)	^{65}Zn; AA^a	[42]
	Co(II)	^{58}Co; AA^a	[35,36,39]
	Fe(III)	^{59}Fe; AA^a	[35,39]
	Cd(II)	AA^a	[39]
	Ni(II)	Spectrophotometry	[39]

of 98.4% after only 30 min [53]. Sukiman examined the uptake
of gold from standard solutions (0.02 to 25 ppm) in 1 M hydrochloric acid [16]. The acidic gold samples were passed through
the glass columns packed with foam ($\sim$0.5 g) or similar masses
pretreated with various organic solvents. The uptake of ^{198}Au
by nonloaded foam from 0.06 ppm samples was 99.7% compared with
99.1% for isopropyl ether/PUF and 99.5% on ethyl acetate/PUF
columns, respectively, whereas 100% uptake was observed for the
25-ppm gold solutions. The absorbed gold could be removed from
unloaded and solvent-loaded foams to the extent of 97.8 to
102.6% using acetone or warm ($\sim$ 50°C) 0.3 M thiourea/1 M hydrochloric acid. Moreover, the foams could be reused several times
without loss of capacity provided columns were first washed with
1 M hydrochloric acid [16]. The technique was successfully
extended to extract gold from natural waters (Whiteknights Lake
on Reading University campus).

Braun and Farag also examined the recovery of gold as its
thiourea complex, $Au(thiourea)_2^+$, in perchloric acid solution
on five different polyether foams and one polyester foam. The
uptake of the cationic complex varied from 85.2% to 93.2% for
the open-cell polyether-type foams compared with 77.2% for the
polyester sample. The absorption by the polyether foams related
partly to cell dimension and decreased as cell size decreased.
The sorption by 10 g of foam material from 0.1 M perchloric acid
(1 dm^3) containing 3% thiourea, 1% sodium perchlorate, and 1 mg
gold(III) approximately matched that absorbed by just 0.2 g of
active carbon under comparable conditions, but foam is cheap
and unlike carbon eliminates the need for tedious filtration.
The rate of sorption on foams (expressed as $t_{\frac{1}{2}}$) varied from
1.3 to 2.0 min compared with 0.6 min for active carbon [20].

Aqueous irridium(IV) is reduced to Ir(III) in the presence
of foam. The reduced species is not significantly absorbed by
polyether-based foams; hence Moore and Chow investigated the

sorption of hexachloroiridate(IV) from organic solvents [46].
Distribution ratios for acetone and ethyl acetate were 225 and
$1.1 \times 10^4 \ dm^3 \ kg^{-1}$, respectively. The rate of sorption from
acetone was enhanced by adding acetic or trichloroacetic acid
but decreased with added water, probably due to formation of
nonsorbable iridium(III) species by hydrolysis. In fact, up
to 93% or iridium(IV) could be stripped from the iridium-loaded
foams after reduction to iridium(III) with 5 M hydrochloric acid
within 40 min, but recovery was still incomplete, even after 30
hr [46]. The behavior of iridium(IV) and rhodium(III) has also
been studied [113] using silicone rubber foam (see Secs. 1.3.1
and 2.3).

Hexachloroplatinate(IV) was also extracted from acetone,
ethanol, and propan-2-ol, but the greatest effect was seen with
ethyl acetate (D $\sim 4.8 \times 10^3$) [46].

[A recent separation scheme using a polyether foam was based
on the fact that in 2 M hydrochloric acid the extraction of
iridium(IV) decreased as the potassium thiocyanate was decreased
below 10^{-2} M, whereas the extraction of rhodium(III) increased
[Al-Bazi and Chow, *Anal. Chem.*, 53:1073 (1981)]. Thus, for a
mixture comprising rhodium (15 ppm) and iridium (200 ppm) in
potassium thiocyanate (2×10^{-3} M) and 2 M hydrochloric acid,
85.92% of the rhodium(III) was retained by the foam compared to
only 8.99% for the iridium(IV).]

2.1.2 Antimony [12,13,111]

The recovery of antimony, which is rather toxic, and found in
many commercial products, is of interest.

Antimony(III) and (V) samples containing variable amounts
of lithium chloride or hydrochloric acid were loaded onto each
foam using an automatic squeezing device with an eccentric cam
[111]. The squeezer depressed and released up to 10 glass
plungers (24 times per minute) in 10 glass cells containing

TABLE 3 Effect of Lithium Chloride on the Sorption of Antimony
in 0.7 M Hydrochloric Acid by Two Different Open-Cell Polyether
Foams

	Distribution coefficients ($dm^3 kg^{-1}$)		
	Regular foam	Acid-resistant foam	
LiCl (M)	Sb(III)	Sb(III)	Sb(V)
4	6 (2)[a]	3 (1)	3 (1)
6	--	398 (57)	414 (58)
7	4360 (93)	6150 (95.3)	10414 (97.2)
8	--	2200 (88)	14700 (98)
10	95 (21)	398 (57)	37206 (99.2)

[a]Parenthesized values are percentage extractions.
Source: Ref. 111.

foam and kept in an air thermostat at 22 ± 0.5°C. The uptake of
antimony(III) and (V) species was influenced by the lithium
chloride concentrations at 0.7 M hydrochloric acid (Table 3).
Hydrogen ion concentration is also critical, e.g., at a fixed
background level of 7 M lithium chloride, the percentage extraction of antimony(III) on the acid-resistant polyether foam was
very poor in 0.1 M hydrochloric acid compared to 85% and 95.3%
in 0.5 M and 0.7 M hydrochloric acid, respectively. The comments
on the disparity for D values of 75 to 500 recorded for antimony(V)
by Bowen [12] and 37,200 ± 400 in this work [111] are unrealistic
since the foams could be different, and moreover, lithium chloride
was absent.

Recoveries of each antimony species were measured by counting
either the foam activity or the various stripping agencies. Direct
counts of [124]Sb on the foam indicated 99.7 ± 0.3% recovery for
antimony(III) and 99.9 ± 0.3% for antimony(V) after acetone
stripping. The [124]Sb levels in the acetone stripping solvent
indicated 100% recovery of antimony(III), but other strippers

were less efficient, e.g., 21% for 0.1 M hydrochloric acid, 90%
for water, and 91% for 0.1 M sodium hydroxide [111].

2.1.3 Tin [57]

Unlike antimony, tin is relatively nontoxic but still considered
a pollutant, and Lo and Chow recently investigated the extraction
of tin(IV) in hydrochloric, hydrobromic, hydrofluoric, and hydro-
iodic acids, nine different group I halides, and seven miscel-
laneous chlorides using three different foams (Table 4). Preliminary batch and flow systems indicated maximum uptake of
tin(II) and tin(IV) by the polyester from ∿6 M hydrochloric
acid. However, as recorded elsewhere, polyesters are acid
labile [36] and in this work disintegrated after about 45 min
exposure to 6 M hydrochloric acid and studies were confined to
the polyether-type foams and tin(IV).

The D values indicate that hydrochloric acid/lithium
chloride is superior to acid alone for the three different foams
(Table 4). Recovery of tin(IV) by 0.1 M hydrochloric acid was
95 ± 3% for both types of loaded polyether foams and 98.7 ± 0.5%
for regular foam using acetone stripper. Precision was improved
by counting the ^{124}Sb species on the foam rather than the recovery
medium [57].

TABLE 4 Distribution Coefficients for Tin(IV) on Three
Different Polyurethane Foams

Foam type	Distribution Coefficient (D)	
	HCl only	HCl (0.12 M) + LiCl (10 M)
Regular polyether	141 ± 9 (3)[a]	5630 ± 119
Acid-resistant polyether	129 ± 7 (4)	2280 ± 48
DiSPo polyester	120 ± 8 (6)	732 ± 23

[a]Parenthesized values are hydrochloric acid molarities.
Source: Ref. 57.

2.1.4 Gallium(III) and Iron(III) [47,48,114,115]

During studies conducted with tantalum tailings, it was noticed
that gallium in acid media could be extracted by open-cell polyether type foams and, moreover, without interference from greater
than 1000-fold amounts of aluminium [47]. However, iron(III)
absorbed in a foam affects the efficiency for gallium(III)
absorption, and when extracting gallium(III) from gallium/iron
mixtures, some pretreatment would be necessary. Leaching of
ores is generally effected with sulfuric acid, but gallium(III)
is not extracted in 5 M sulfuric acid. However, addition of
sodium chloride or hydrochloric acid (~8 M) raises the extraction to ~90%.

Dynamic studies on foam columns showed that the efficiency
of gallium extraction in acidic solution, as well as its recovery
by water, depended on temperature and flow rate; both uptake and
recovery increased with decreasing flow rates. The extraction
efficiency can be enhanced by passing either the gallium sample
through several foam plugs or the same solution through the foam
several times. Under optimal conditions, gallium loads of ~ 10%
m/m were achieved. Ground, closed-pore, rigid polyurethane foam
showed similar extraction profiles but with lower efficiency
[47]. Gallium can also be completely recovered from foams using
2 M sodium hydroxide washes [48].

2.1.5 Metal-Thiocyanato Systems

Among the substances showing negligible absorption on many types
of polyurethane foams, Bowen listed ammonium phosphate, copper(I)
and (II) acetates, nickel nitrate, palladium chloride, and silver
nitrate [12]. Group II chlorides [60], zinc nitrate [60], lead
nitrate [60], manganese chloride [60], and cobalt(II) chloride
[36,60] behave similarly. Relatively small quantities of nickel
(~5%) are absorbed between pH 0.9 and 11 [52] compared to copper
[54] (~40%), while cadmium showed a strong affinity for an open-
pore polyurethane foam [54].

Bowen also reported that polyether foams could extract
metal complexes, especially with chloride and thiocyanate
ligands [12]. Thus, further studies have been directed toward
the behavior of transition metal and group IIB metal thiocyanato-N
and -S complexes toward polyurethane and foams [35,36,38-40,42,58,
60,116,117]. Thiocyanato complexes of nickel [39], lead [60],
and manganese [60] show little affinity for polyether-type polyurethane foams. However, at high-level backgrounds of potassium
chloride, this trend can be reversed, at least for manganese
(Fig. 1). The effect is possibly related to ionic strength
[117]. Considerable fractions of other metal thiocyanate complexes are extracted, and the normal white polyether foam becomes
colored red and blue, respectively, after contact with iron(III)
and cobalt(II) thiocyanate mixtures [35,60]. On the contrary,
neither of these complexes can be extracted by PPI-800nFR polyester foams [35]. The extraction efficiencies for iron(III) and
cobalt(II) at the 1-μg dm^{-3} level was 96.1 ± 0.6% and 95.9 ± 0.8%,
respectively [35], on unloaded polyether foam columns at flow
rates of 30 cm^3 min^{-1}. More or less complete extraction was
possible, even at 40 cm^3 min^{-1}. Flow rates have also been specified in units of volume (area^{-1} per unit time), so as to include
the cross-sectional area of the column which thus has the dimensions of a linear flow rate [118].

The effect of temperature on the extraction is interesting.
The cobalt(II) extraction fell by 5% on raising the temperature
from 22 to 31°C, while that of iron(III) fell by 12% over the
same range. At higher temperatures, still the extraction fell
drastically and at ∿70°C the foam becomes white again due to
breakdown of the colored complexes. The reaction is, however,
reversible, and the foam reverts to its blue or red color on
cooling the foam-solution system [39]. The intensity of color
produced by such foam-metal complex interactions has been used
in semiquantitative assays of metal ions [26].

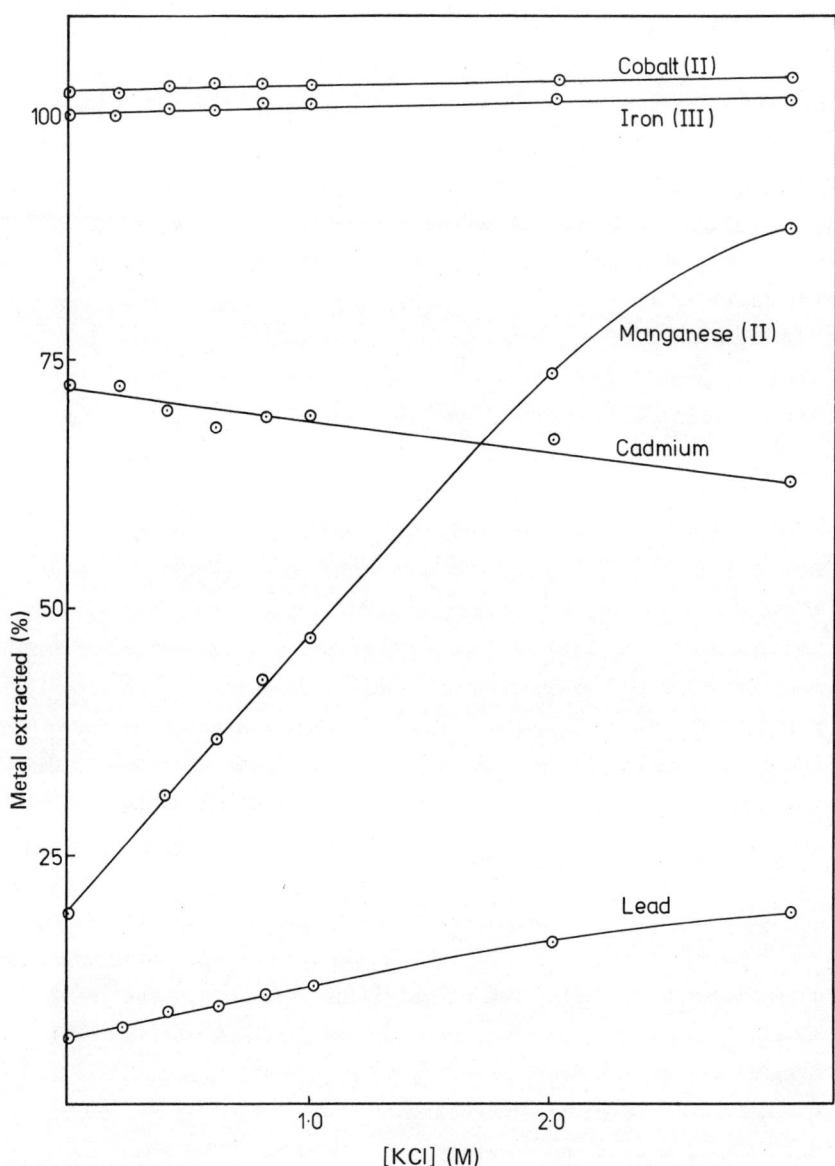

FIGURE 1 The effect of potassium chloride on the extraction of metal ions from 0.6 M potassium thiocyanate solutions.

Thiocyanato complexes of zinc, cobalt(II), and iron(III)
can also be desorbed from polyether columns with hot 0.1 M
nitric acid, indicating their scope as a preconcentration technique for such cations [39]. Acetone will elute 64 and 35%,
respectively, of sorbed cobalt(II) compared to 82 and 77% for
iron(III), whereas neither system can be eluted with diethyl
ether. The capacity of the foam for iron(III) (22 g kg^{-1}) was
twice that of cobalt(II) [39].

The extraction of zinc, mercury(II), and indium(III) is
highly dependent on thiocyanate concentration and foam type.
Thus, D values for indium(III) on a polyether foam increased
from 3717 to 17,499 dm^3 kg^{-1} compared with 743 to 2088 on a
Eurofoam polyester under the same conditions, namely, at potassium thiocyanate levels from 0.1 to 1 M and pH range of 2 to 3.
Extractions were essentially the same over a wide pH range for
any one foam [42]. Absorption is very rapid, e.g., 95.3% of
the indium(III) and 99.8% of the zinc were extracted on unloaded
polyether foam after one minute in batch experiments. Zinc was
recovered from the foams using 1 M nitric acid (150 cm^3) at a
flow rate of 2 cm^3 min^{-1}. Mercury(II) and indium(III) were
stripped [42] with 80% acetone/1 M potassium thiocyanate at
15 cm^3 min^{-1}.

The fact that zinc and indium(III) are sorbed by a polyester foam [42], whereas cobalt(II) and iron(III) are not sorbed
[35], is not paradoxical but a reflection of their structural
differences and is relevant to the mechanistic aspects of foam-
iron interactions to be discussed in Sec. 2.1.6. Comparisons
between one metal-foam system and another are difficult because
the nature of the foam material in published reports is, with
few exceptions [52,60,111], omitted in the experimental content.

The quantitative extraction of metal ions by foams has been
evaluated in many instances with radiotracers (Tables 1 and 2),
and the effect, if any, of γ radiation on foam-metal interactions

is therefore of interest. Thus, samples (0.1 g) of Dunlopillo
polyether foam (D2 code) were irradiated for various times in a
^{60}Co-gamma source (1.6 × 10^{-2} J cm^3 min^{-1}). The capacity of
each irradiated foam for both cobalt(II) and iron(III) thiocyanate complexes was measured and any physical changes noted.

Irradiation periods up to 36 days had little effect on the
sorptive properties of the foams (Table 5). However, the same
is not true of its physical properties, and samples irradiated
for longer than 7 days showed progressive signs of breakdown by
fragmentation into fine particles, a condition which offsets
the favorable hydrodynamic properties of foams. Schnecko and
Bieber [11] used ground-up foam, in which case the above kind
of radiation damage would not be too significant, even though
foams might disintegrate during the equilibrium shaking or
pressing/releasing cycle operations.

2.1.6 Mechanistic Aspects of Interactions between Aqueous Inorganic Species and Unloaded Polyurethane Foams

General Aspects

As mentioned earlier, Bowen noted the resemblance between
the extraction behavior of substances absorbed on polyurethane
foams and that of diethyl ether [12]. This supported the involvement of absorption rather than adsorption, but adsorption of a
monolayer of 1-[^{14}C]stearate was nonetheless assumed for the
purpose of surface area measurements [110]. Bowen also proposed
that nitrogen atoms in the urethane skeleton might function as
weak anion-exchangers or that protonation of oxygen atoms in
polyether linkages could provide alternative reaction sites [12].

This simple absorption concept has found general acceptance
and is featured in subsequent reports concerning a mechanistic
approach to metal-foam interactions. Thus, Braun and Farag
stated that unloaded polyether-type foams can act as solid ether
solvents for the extraction of thiocyanato complexes of cobalt(II)
and iron(III), whereas their polyester counterparts, e.g.,

TABLE 5 Capacity of Irradiated Polyether Foam for
Cobalt(II) and Iron(III) from 0.5 M Potassium Thiocyanate
(Initial [Metal] = 200 ppm)

Irradiation time (d)	Extraction (%)	
	Co(II)	Fe(III)
0	31	57
6.7	29	55
11.4	31	59
21.7	33	59
29.5	33	57
36.2	24	55

Source: Refs. 39, 40.

PPI-800nFR, failed to extract either species under comparable
conditions [35]. Lo and Chow proposed that species such as
$HSbCl_4$ and $HSbCl_6$ may be involved in the extraction of antimony
(III) and (V) by polyurethane foams [111].

A solid oxypropene/oxyethene-based polyether foam has been
equated to a liquid extractant of a moderate permittivity for
the iron(III) system [115]. The variation in distribution
coefficients with high and low hydrogen ion concentrations in
the aqueous phase was related mainly to the extracted species
$HFeCl_4$ and $FeCl_3$, respectively. Electron paramagnetic resonance
spectroscopy of the iron(III)-loaded polyether foams also suggested that either $HFeCl_4$, or the ion pair $H^+FeCl_4^-$, is dissolved
in the solid foam support. The uptake of hydrochloric acid from
an iron(III) free solution was 5 ± 1%, whereas no lithium
chloride uptake was evident under similar conditions [115].

A similar theme is introduced in studies of tin(IV) where
the $HSnCl_5$ molecule formed in the following reactions:

$$Sn^{4+} + Cl^- \rightleftharpoons SnCl^{3+} \tag{1}$$
$$SnCl^{3+} + Cl^- \rightleftharpoons SnCl_2^{2+} \tag{2}$$

$$SnCl_2^{2+} + Cl^- \rightleftharpoons SnCl_3^+ \tag{3}$$

$$SnCl_3^+ + Cl^- \rightleftharpoons SnCl_4 \tag{4}$$

$$SnCl_4 + Cl^- \rightleftharpoons SnCl_5^- \tag{5}$$

$$SnCl_5^- + H^+ \rightleftharpoons HSnCl_5 \text{ (aq)} \tag{6}$$

could be a major component in the distribution mechanism

$$HSnCl_5 \text{ (aq)} \rightleftharpoons HSnCl_5 \text{ (foam)} \tag{7}$$

and at low pH even H_2SnCl_6 may be formed [57]. For an acid-resistant polyether Union Carbide foam, the capacity for tin(IV) was 62 ± 8 mg g^{-1} foam, which demands a surface area of about 111 m^2 g^{-1} for a viable adsorption mechanism against an experimental figure of 0.081 m^2 g^{-1}. Alternatively, the extraction could involve protonation of ether sites [57].

An essentially identical set of equilibria is invoked [48] for the gallium(III)-polyether foam system, wherein the equilibrium reaction which most readily explains the extraction is

$$HGaCl_4 \text{ (aq)} \rightleftharpoons HGaCl_4 \text{ (foam)} \tag{8}$$

As with the tin(IV) study [57], the capacity of the polyether foam was again affected by $[H^+]$, $[Ga^{3+}]$, and $[Cl^-]$ in all the equilibria leading to the final extractable product:

$$H^+ + GaCl_4^- \rightleftharpoons HGaCl_4 \text{ (aq)} \tag{9}$$

Mechanism I involving reactions listed above can be replaced, or supplemented, in mechanism II by

$$H^+ + \text{---}CH_2 - O - CH_2\text{---} \rightleftharpoons \text{---}CH_2 - \overset{H}{\underset{+}{O}} - CH_2\text{---} \tag{10}$$

followed by

$$\text{---}CH_2 - \overset{H}{\underset{+}{O}} - CH_2\text{---} + GaCl_4^- \rightleftharpoons \text{---}CH_2 - \overset{H}{\underset{GaCl_4}{O}} - CH_2\text{---} \tag{11}$$

Gesser and Horsfall could not distinguish between these two mechanisms but emphasized that the capacity of the foam ($\sim$10% m/m) was not accountable in terms of surface adsorption

by a foam with a surface area of 81 $m^2 g^{-1}$, as found by the BET
technique using krypton [48]. Moreover, extraction was enhanced
by raising the [LiCl] or [HCl] levels and in the latter case
could equate to the extent of the protonated sites, Eq. (10).

Diffusion experiments have been conducted with a two-cell
system separated by a thin membrane comprising polyether foam:
one cell containing just water and the other gallium(III),
gallium(III) with hydrochloric acid, or gallium(III)/lithium
chloride/hydrochloric acid [48]. Diffusion of gallium(III) from
a 3 M HCl/3 M LiCl medium was faster than from just 3.12 M HCl
but slower than for 6.07 M HCl alone. After 230 hr atomic
absorption spectroscopic analysis of the cells showed that
virtually all the gallium(III), but none of the lithium, had
diffused across the intact polyether film which itself had
developed blisters and vacuoles.

This active transport of gallium(III) is again explicable
in terms of either of the above mechanisms [48]. In the first
case, the complex $HGaCl_4$ is formed in the acid-loaded cell,
followed by dissolution in and diffusion across the thin membrane into the water cell, where it hydrolyzes to various ionic
species which cannot back diffuse. In the second scheme, the
membrane is first protonated, Eq. (10) followed by uptake of the
tetrachlorogallate(III) complex which then jumps from one protonated ether site to another down a concentration gradient to
the water-only medium. Here again, hydrolysis produces ionic
gallium species which cannot return [48,114]. The diffusion of
complexes such as $GaCl_4^-$ [48,114], $FeCl_4^-$ [114], $Co(NCS)_4^{2-}$
[119], and UO_2^{2+} [114], respectively, through intact foams is
considered to relate to true absorption into the polymer bulk
[116].

The behavior of hexachloroiridate(IV) in acetone or ethyl
acetate is not compatible with a simple solvent extraction
mechanism and where D values remained essentially independent,

namely, at 225 dm^3 kg^{-1} between 8.6×10^{-5} M and 5.1×10^{-5} M iridium(IV), they fell as the iridium(IV) was raised [46]. The following equilibria have been proposed for the iridium(IV)-polyether foam system:

$$H_2IrCl_6\,(foam) \quad \underset{}{\overset{K_E}{\rightleftharpoons}} \quad H_2IrCl_6\,(org)$$

$$\Big\Updownarrow K_F \qquad\qquad\qquad\qquad \Big\Updownarrow K_{org} \qquad (12)$$

$$H^+(foam) + HIrCl_6^-\,(foam) \qquad H^+(org) + HIrCl_6^-\,(org)$$

Moore and Chow [46] employed this model, where H_2IrCl_6 represents the undissociated ion-association species, to derive an expression for the distribution coefficient D in terms of the extraction coefficient K_E of the undissociated complex:

$$D = \frac{K_E[H_2IrCl_6]^{0.5}_{(org)} + K_E^{0.5}K_F^{0.5}}{[H_2IrCl_6]^{0.5}_{(org)} + K_{org}^{0.5}}$$

Thus, if both K_F and K_{org} approach 0, then D will be constant and equal to K_E; whereas if only K_{org} approaches 0, or if $K_F \gg K_{org}$, then D will decrease with increasing $[H_2IrCl_6]_{(org)}$ as indeed is observed. The permittivity of acetone ($\sim$21) is expected to exceed that of foam, and K_F is thus unlikely to be greater than K_{org} when extracting with acetone, but in the case of ethyl acetate ($\varepsilon \sim 6$), K_F might approach K_{org}. Moore and Chow concluded from the analysis of their data that the variation in D values are inexplicable in terms of just solvent extraction, and they suggested the possibility of specific foam sites capable of absorbing iridium species [46].

To date, surface area measurements and diffusion studies have established the concept of true absorption, and metal adsorption can be rejected [116]. The involvement of ion exchange is not incompatible with the weak base anion capacity for polyurethane foams reported by Navratil and Sievers [9]. The extraction of thiocyanato complexes of cobalt(II) [116] and

iron(III) [39] falls very sharply with increasing acidity whereas
the reverse trend might be expected if protonation occurred at
nitrogen atom sites in the urea, allophanate, biuret, or urethane
linkages or at the ether or ester units. All such sites are
rather weakly basic and cannot account for the high metal-foam
capacities. Hamon et al. have concluded that any anion-exchange
contribution must arise at sites generated by some other mechanism [116]. On the other hand, it is worth noting that the
uptake of gallium(III) [48] and iron(III) [115] in chloride
systems increases as the concentration of hydrogen is raised.

   The loan pairs on the nitrogen and oxygen atoms have also
been considered in terms of polyurethane ligands (>L:). Complexation could arise by addition,

$$2 \text{>L:} + \text{Co(NCS)}_4^{2-} \rightleftharpoons \text{Co(NCS)}_4(\text{:L<})_2^{2-} \tag{13}$$

or by substitution

$$\text{>L:} + \text{Co(NCS)}_4^{2-} \rightleftharpoons \text{Co(NCS)}_3(\text{:L<})^- + \text{SCN}^- \tag{14}$$

The change from tetrahedral to octahedral symmetry associated
with reaction (13) would be accompanied by a large change in
the electronic spectrum, namely, a blue-to-pink color change
[116]. Visually, however, such cobalt(II)-loaded foams are
blue-green and possess a visible spectrum characteristic of the
tetrahedral Co(NCS)_4^{2-} species. The colors of other metal complexes on polyurethane foams are also typical of their respec-
tive tetrahedral species, e.g., Pd(SCN)_4^{2-} and FeCl_4^-. Chemical
analysis of iron(III) chloride foams gave [12] a formula corre-
sponding to $\text{FeCl}_{3.85}$, and further attention has been recently
given to the stoichiometries of transition metal thiocyanate
complexes absorbed on polyether foams [60].

   Complexation by ligand exchange should also effect some
change in the visible spectrum, but of course no such alteration
is evident, and, moreover, the spectrum of the complex eluted
from the foam with acetone, methyl isobutylketone, or tetraglyme

is that of aqueous $Co(NCS)_4^{2-}$. A ligand-exchange mechanism is
thus considered unlikely and cannot be commonplace [116]. The
ultraviolet-visible spectrum of the cobalt(II) thiocyanate complex eluted from loaded foam with acetone is closely similar to
that of an authentic $Co(NCS)_4^{2-}$ complex in acetone [60].

The etherlike solvent syndrome has, of course, found wide
acceptance and yet has not been rigorously tested [116]. Indeed,
it becomes less credible on the basis of a comparison among distribution coefficients [48,57,111,115,116,119,120] for simple
ethers and polyurethane foams (Table 6). The numerical disparity
for the distribution coefficients of $Co(NCS)_4^{2-}$ between diethyl
ether and polyurethane foam is truly striking, and the same trend
is also obvious for the other systems listed (Table 6). This
sort of anomaly led Hamon et al. to remark that factors other
than simple etherlike solvent extraction are operative and that
some specific interactions must exist within the polymer [116].

Several clues as to the nature of this special site relationship are available from independent work involving adducts
formed between inorganic salts, especially group I tetraphenyl-
borates, and macrocyclic (crown) ethers, e.g., 18-crown-6 [121,

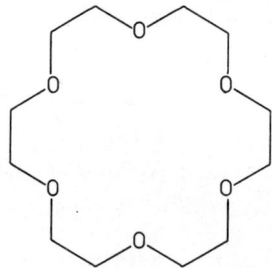

122], as well as acyclic poly(alkeneoxy) systems [123,124] with
a $-CHR-CH_2O-$ repeating unit. They include (in particular regard
to certain polyurethane foams) polyethene oxides (PEOs)

$$HO(CH_2CH_2O)_nH$$

TABLE 6 Distribution Coefficients for Inorganic Complexes by Ethers and Polyurethane Foams

Ion	Phase Aqueous	Organic	D $(dm^3\ kg^{-1})$	Ref.
Sb(III)	6 M HCl	Diethyl ether	0.06	[120]
	6.5-8.5 M HCl	Diisopropylether	0.16	[120]
	7 M HCl	Polyether foam	416	[111]
Sb(V)	6 M HCl	Diethyl ether	4.3	[120]
	6.5-8.5 M HCl	Diisopropylether	199	[120]
	7 M HCl	Polyether foam	500	[111]
Sn(IV)	6 M HCl	Diethyl ether	0.2	[120]
	4 M HCl	Polyether foam	129	[57]
	3 M HCl	Polyether foam	141	[57]
Fe(III)	6 M HCl	Diethyl ether	99	[120]
	7.75-8 M HCl	Diisopropylether	10^3	[120]
	1 M HCl	Polyether foam	10^3	[115]
Ga(III)	6 M HCl	Diethyl ether	32	[120]
	7 M HCl	Diisopropylether	$>10^3$	[120]
	0.85 M HCl	Polyether foam	6300^a	[48]
Co(II)	0.5 M HCl + 1-7 M NH_4SCN	Diethyl ether	0.037-3.03	[120]
	2 M NaSCN + 1 M buffer	Polyether foam	$>10^6$	[119]
	5 M KSCN + 1 M buffer	Polyether foam	$>10^6$	[119]
Pd(II)	0.5 M HCl + 1 M NH_4SCN	Diethyl ether	0.02	[120]
	0.5 M HCl + 7 M NH_4SCN	Diethyl ether	10^{-3}	[120]
	0.5 M HCl + 0.15 M KSCN	Polyether foam	18200	[116]

[a]No measurable extraction of gallium(III) using open-cell silicone rubber sponge.

and polypropene oxides (PPOs)

$$HO(\overset{\overset{\displaystyle CH_3}{|}}{CHCH_2}O)_n H$$

Much of such structures will still be represented in a polyurethane foam after condensation with an isocyanate, as typified
for a PPO-polyol:

$$RNCO + HO - CH_2 - \overset{\overset{\displaystyle |}{CHO}}{\underset{\underset{\displaystyle CH_3}{|}}{}} - CH_2 - \overset{\overset{\displaystyle |}{CHO}}{\underset{\underset{\displaystyle CH_3}{|}}{}} - CH_2 \cdots OH + OCNR$$

$$\longrightarrow RN\overset{\overset{\displaystyle H}{|}}{-}\overset{\overset{\displaystyle O}{||}}{C}- OCH_2 - \overset{\overset{\displaystyle |}{CHO}}{\underset{\underset{\displaystyle CH_3}{|}}{}} - CH_2 - \overset{\overset{\displaystyle |}{CHO}}{\underset{\underset{\displaystyle CH_3}{|}}{}} - CH_2 \cdots \overset{\overset{\displaystyle O}{||}}{O}\overset{\overset{\displaystyle H}{|}}{C}-N-R \qquad (15)$$

Thus, the patterns of extraction of heavy metal thiocyanate
complexes by a D2 polyether foam [39,60], polyethene oxides
[123], and polypropene oxides [123] are rather similar, namely,
Zn $\sim$ Co(II) > Fe(III) > Cd > Pb $\sim$ Ni, while the affinity of the
first three members in this series for the foam is such that
elution with diethyl ether is not possible [39].

X-ray diffraction studies of polyethene oxide/mercury(II)
chloride complexes [125] and of alkali metal ion/cyclic polyether
complexes [122] indicate direct interaction of ether oxygen atoms
with the cations. Such interactions can be expected to influence
the helical structures of the polyalkeneoxy chains and so affect
the infrared spectra. Indeed, the most significant differences
between the reference polyalkeneoxy materials and their tetra-
phenylborate salts concern the carbon-ether-oxygen stretching
frequencies [124]. Thus, for a PEO-barium complex with tetra-
phenylborate, the etheneoxy bands at 1145 and 1060 cm^{-1} have
disappeared, and the main band at 1110 cm^{-1} has shifted to
1085 cm^{-1}. For the analogous barium- and calcium-PPO derivatives the very strong C-O-C stretching frequency of the uncom-
plexed polymer at 1105 cm^{-1} is represented [124] by a very
strong band at 1065 cm^{-1}.

The strong, broad symmetrical C-O-C band at ~ 1100 cm^{-1} in D2 polyether foam remains essentially intact after exposure to cobalt(II) chloride or iron(III) chloride solutions and, in fact, iodine vapor, although in the latter case iodine is sorbed. However, after contact with cobalt(II) or iron(III) chloride solutions containing potassium thiocyanate, the ether vibrational band shifts to ~ 1085 cm^{-1} and is no longer symmetrical, while the very sharp, strong band at 2160 cm^{-1} in a solid Co(NCS)$_4^{2-}$ sample has shifted, but with retention of profile, to 2080 cm^{-1} (Fig. 2). The downward shift of ~ 30 cm^{-1} observed for a similar cobalt(II)-thiocyanate-foam system is regarded as compelling evidence for strong polyether-metal ion involvement [116]. However, the blue-green color of the loaded foam and its visible spectrum, with λ_{max} at 615 nm and a 580 nm shoulder, indicate the uptake Co(NCS)$_4^{2-}$ species with retention of tetrahedral symmetry. Direct interaction of this cobalt(II) species with the ether sites is unlikely, and the infrared shift is interpreted [116] as a result of sodium cation chelation by the polymer in sorbing Na$_2^+$Co(NCS)$_4^{2-}$. Similar trends are reported for group I and group II metal-crown ether- [126] and polyalkylene oxide complexes [124].

The polyether foams nearly saturated with Co(NCS)$_4^{2-}$ exhibited a marked increase in the polymer glass transition temperature, but no mention was made of any restoration of physical properties to the foam after elution of the cobalt(II) complex [116]. Neither of the D2 polyether foams loaded with thiocyanato complexes of cobalt(II) at 4.74 g kg^{-1}, or iron(III) at 12.46 g kg^{-1}, showed any sign of brittleness [60].

Cation Chelation Mechanism [116]

In essence, the cation chelation mechanism (CCM) proposes that cations M^{n+} can be solvated by some degree of selective chelation in a portion of the polyurethane foam,

$$M^{n+} + \overline{Site} \rightleftharpoons \overline{[M\ Site]}^{n+} \tag{16}$$

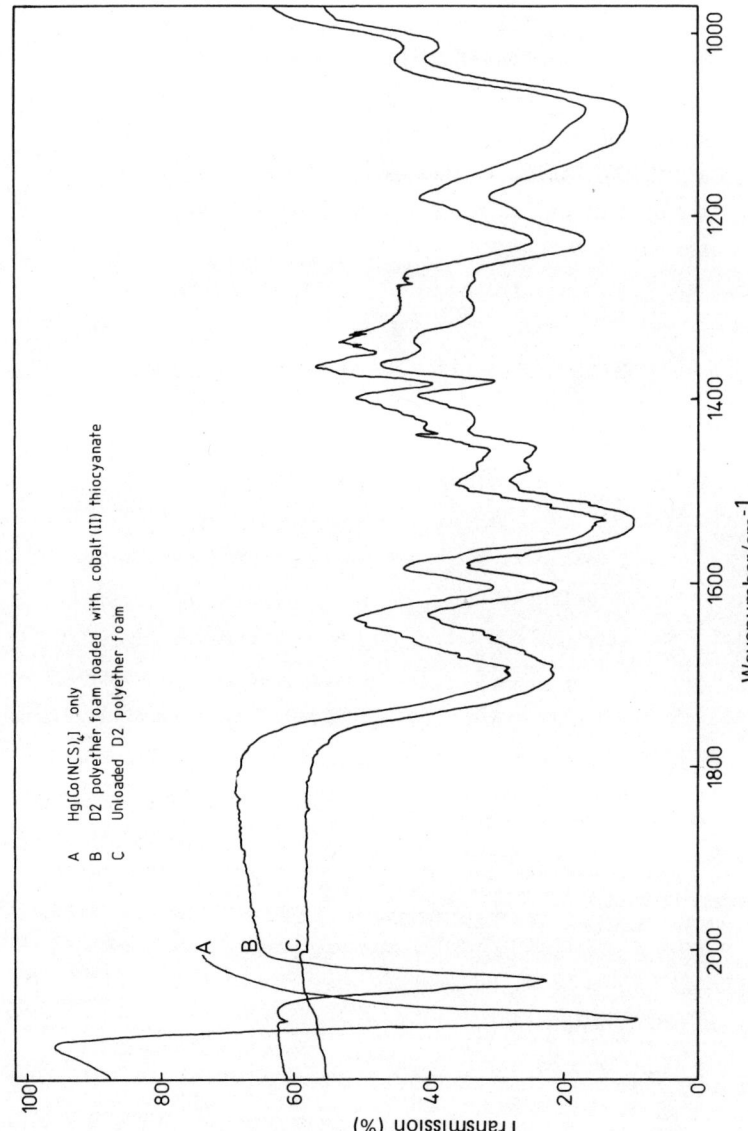

FIGURE 2 Infrared spectra of various polyether foams and mercury(II) N-tetrathiocyanato cobaltate(II) (1 mg % KBr disks).

which can then interact with the metal anion complex A^{n-} to maintain charge neutrality:

$$[\overline{M\ Site}]^{n+} + A^{n-} \rightleftharpoons [\overline{M\ Site}]A \qquad (17)$$

This mechanism by which anion metal complexes are sorbed does not necessarily require protonation of foam sites but is nonetheless related to the weak-base anion-exchange concept [116].

The type of polyol in the foam will exert a critical influence on the actual extent of cation extraction and thus of anion. This is dramatically illustrated by the extraction behavior of $Co(NCS)_4^{2-}$ on seven different types of polyurethane foams under identical conditions (Table 7). The first three foams differ *only* in the ratio of polyethene oxide (PEO) to polypropene oxide (PPO). The stepwise replacement of PPO by PEO groups effects about a 20-fold increase in the distribution coefficients and underlines the chelating superiority of the PEO units. This is probably related to the steric effect of the CH_3 group rather than to differences in electron densities on the respective

TABLE 7   Sorption of Tetrathiocyanato-N-cobaltate(II) by Seven Different Polyurethane Foams

Foam code	Polymer composition	$D\ (dm^3 kg^{-1})$
27CGS-44-2A	100% PPO polyether	1,047
27CGS-44-1	8% PEO polyether 92% PPO polyether	7,260
27CGS-44-3	14% PEO polyether 86% PPO polyether	21,600
A	60% PEO/PPO polyether copolymer 40% Styrene-acrylonitrile copolymer	7,260
B	80% PEO/PPO polyether copolymer 20% Styrene-acrylonitrile copolymer	8,120
1338 BFG	Unknown polyether	16,800
diSPo	Unknown polyester	6

Source: Ref. 116.

oxygen atoms since the electron donating CH_3 group in the PPO
units should enhance their basicity and hence chelating strength
[116]. The greater tendency of polyethene oxides to form helical
structures with inwardly directed oxygen atoms compared to material containing the corresponding PPO units reflects this steric
influence [127]. The same polyether effect is also apparent,
albeit to a lesser extent, in foams A and B (Table 7).

The extremely small D value for the diSPo polyester foam
is attributed partly to the inherent inability of polyesters to
take a helical orientation about a central axis, and it behaves
much like a simple solvent [116]. However, it will be recalled
that antimony was absorbed to a comparable extent on both a
polyether and a polyester foam (Table 4), but again the type of
polyol unit was not specified. It is just this sort of crucial
information on the chemical nature of the polyol component in
commercial foams which is absent in many published studies on
metal-ion foam interactions.

Cation and anion selectivity measurements have been made
to further support the CCM concept [116]. Pedersen [121]
reported the apparent order of stability for 18-crown-6 and its
derivatives to be

$$Li^+ < Na^+ < Cs^+ < Rb^+ < K^+$$

This selectivity order was extended to the "A" isomer of
dicyclohexyl-18-crown-6; thus

$$Li^+ < Cs^+ < Na^+ < Rb^+ < K^+ < Ag^+ < Tl^+$$

and

$$Ca^{2+} < Hg^{2+} < Sr^{2+} < Ba^{2+} < Pb^{2+}$$

for monovalent and divalent cations, respectively [128]. The
order of affinity for a polyether-type foam for various monovalent and divalent cations [116] was as follows:

$$Li^+ < Na^+ < Cs^+ < Rb^+ < K^+ \sim NH_4^+ < Ag^+ < Tl^+$$

and

$$Ca^{2+} < Ba^{2+} < Hg^{2+} < Pb^{2+}$$

The close similarities between these extraction orders [116] and those of crown ethers [121,128,129] and polyalkene oxides [129], for example, strongly support the cation chelation mechanism.

The efficient solvation of cations by a foam facilitate the extraction of anions, and the most readily extracted anion species partly relate with Hofmeister series:

$$F^- < Cl^- < Br^- < I^- < NCS^-$$

The extraction of potassium, thallium(I), and silver also depends strongly on the anion present in the solution contacting the foam.   For a polyether foam cation sorption was found to increase in the order:

$$NO_3^- < 2,4\text{- and } 2,6\text{-dinitrophenolate} < picrate < 8\text{-anilino-}$$
$$1\text{-naphthalene sulfonate} < tetraphenylborate$$

and thereby matching both increasing size and polarizability of the anion [116].

The various investigations outlined above, particularly for the sorption of $Co(NCS)_4^{2-}$ by foams of known formulation, strongly supports this CCM proposal [116], and it is beholden of authors in future publications to report the fullest struc-tural details of the actual polyurethanes employed.

2.2   LOADED FOAMS

Unloaded foams offer a simple means of rapidly sorbing reasonable amounts of ionic materials from aqueous media but with a limited degree of selectivity.   Chelates, on the other hand, e.g., dimethylglyoxime at pH > 7, exhibit an exclusive preference for nickel, and the utility of foams comprising such metal-specific agencies was briefly described by Bowen [13].   These were

conveniently produced by incorporating the chelate into the
isocyanate/polyol reaction mixture. Mercury(II) could be
extracted from aqueous solutions using foams with 1% m/m of
dimethylglyoxime, phenylthiourea, 4-methyl-2-thiouracil,
sulfadiazine, or sulfathiazole, while silver was similarly
sorbed by the phenyl thiourea-loaded foam [13].

In 1972 Braun and Farag described the analytical use of
open-cell polyurethane foams impregnated with tri-n-butyl
phosphate as the stationary phase [17]. Since then, foams
loaded with other extractants, chelates as well as organic
ion-exchangers and inorganic precipitates, have been successfully employed for various analytical schemes. Such materials
may also be simply loaded by adding the stationary phase to the
actual polyurethane foam sample.

The polyurethane foam seems to function as a relatively
inert support on which high loadings can be superimposed, e.g.,
50 to 60% m/m, and yet without any serious impairment of the
normal favorable hydrodynamic character which is so obviously
an asset in the art column chromatography. Thus, relatively
high flow rates, e.g., 50 to 60 cm^3 min^{-1}, are readily attained
simply by gravity flow, and moreover, polyurethane foams retain
a higher amount of organic support than other presently available
granular supports such as Kel-F [17].

In these systems it seems reasonable to relate the mechanistic aspects to interaction(s) between species in the solution
system and the loaded material and *not* to the polyurethane foam
support. However, the infrared features of a D2 polyether foam
are retained when loaded with diethyl hexyl phosphoric acid
[60]. It is also interesting to note that the uptake of tri-
butylphosphate by five different types of open-cell polyether
foams ranged from 65.4 to 68.5% m/m compared to 50.7% m/m for
the open-cell PPI-800nFR polyester foam [20].

Most of the loaded reagents have been incorporated into
the final commercial foam product by simple, direct physical
immobilization, but foams have also been synthesized with
specific functional groups [56,92].

The hydrophobic character of polyurethane foams, together
with relatively high surface areas, allows the immobilization
of considerable amounts of a variety of organic reagents. Thus,
tri-n-butyl phosphate is readily incorporated into a foam matrix
by simply stirring together for a few hours. In this case the
polyurethane foam is considered [18] as a support for the tri-n-
butyl phosphate which is actually the stationary phase. Separation using such loaded foams is called *reverse-phase foam chromatography* [17,20-22,25].

2.2.1 Foam-Supported Solvent Extractants

Tri-n-butyl phosphate, $(C_4H_9)_3P=O$ (TBP), is a well-established
liquid extractant for a variety of metal ions and one of the
first solvent extractants to be supported on polyurethane foams
by Braun and Farag [17] for batch and column work (Table 8). A
simple vacuum procedure has been devised to avoid air bubbles
during homogeneous packing of glass columns (typically, 10 and
15 mm diameter by 15 cm long) with foam [17,21] and other supports, e.g., Voltalef. In some cases, experiments at controlled
temperature were undertaken [21] with water-jacketed columns at
±0.1°C.

Preliminary experiments on the extraction of palladium
with TBP-loaded polyurethane proved more suitable than either
poly(vinyl chloride) or viscose rubber and loadings of TBP were
about twice that on Kel-F [17]. The rate of extraction of the
palladium-thiourea complex was fast as shown by a batch technique when $t_{\frac{1}{2}}$ for equilibrium adsorption was 0.5 min [21].

In column work the narrow, yellow thiourea complex band
retained just on top of the column was quantitatively eluted

TABLE 8 Selected Studies of Metal Ion/Tri-n-butyl Phosphate-Loaded Polyurethane Foam Systems

Source and type of foam	General pretreatment and foam loading technique	TBP (% m/m)	Typical sorption system		Analysis technique	Ref.
Polyether D2 (Dunlop)	A.R. HCl (1 M) for 1 hr. Washed in deionized water, then acetone. Stirred with TBP for several hours. Excess TBP removed by pressing between filter papers.	48.5 and 54.1	Thiourea 3% m/v, NaClO$_4$ 1% m/v, HClO$_4$ 0.1 M	Bi + Ni	Spectrophotometry	[39]
Open-cell polyether foams (N. Hungarian Works, Sajóbábony)	Washed in acetone. Dried. Cubes or cylinders equilibrated with TBP (3 cm^3 g^{-1} foam). Dried between filter paper sheets.	--	Thiourea 3% m/v, NaClO$_4$ 1% m/v, HClO$_4$ 0.1 M	Pd + Ni	Spectrophotometry	[17]
		--	Thiourea 6% m/v, NaClO$_4$ 2% m/v, HClO$_4$ 0.2 M	Pd + Ni, Pd + Bi, Pd + Bi + Ni	Spectrophotometry	[21]
	Not given		Up to 8.4 M HCl for batch work. 10^{-2} M or 4 M HCl for columns	Fe(III); Co(II); Ni; Cu	Compleximetry and 58Co, 59Fe	[22]
	As in ref. 21.	~67	Thiourea 6% m/v, NaClO$_4$ 2% m/v, HClO$_4$ 0.2 M	Au(III) in various mixtures of Zn; Co(II); Ni; Fe(III); Cu; Sb(III); Bi; Pd and Ag	198Au	[25]
Six different foams (see Table 9)		50.7 → 68.4	Thiourea 3% m/v, NaClO$_4$ 1% m/v, HClO$_4$ 0.1 M	Au(III)	198Au	[20]

with water at 1 cm^3 min^{-1}, the elution curve being sharp and
symmetrical [21]. The height equivalent to a theoretical plate
(HETP) was calculated from

$$N = \frac{8V_{max}^2}{W^2} = \frac{L}{HETP}$$

where

> N = number of plates
>
> V_{max} = volume of eluate to peak maximum
>
> W = width of peak at $1/e$ times maximum solute
>   concentration
>
> L = length of the foam bed

The HETP value of 1.7 mm ± 7% was also confirmed from a
breakthrough capacity curve calculation:

$$N = \frac{\overline{V}V}{(\overline{V} - V)^2}$$

where

> $\overline{V}$ = volume of effluent at center of S-shaped breakthrough
>   curve when [Pd(II)] is one-half its initial concentra-
>   tion
>
> V = volume at which effluent has concentration 0.1587 of
>   initial concentration

The increased plate height (that is, decrease in column
performance) with increasing flow rates, ranging between 0.9
and 6.4 cm^3 min^{-1}, was only pronounced at 25°C. On the other
hand a slight increase in plate height with increased flow rate
occurred at 35°C and 45°C so that a relatively high flow rate
could be applied without any appreciable loss in column performance [21]. The slightly lower value of ∿1.4 mm reported in preliminary work [17] could relate to differences in flow rate.
The HETP value for TBP-loaded Voltalef powder was 2.8 mm.

The *breakthrough capacity*, defined [21] as the amount of
palladium-thiourea complex that could be retained on the column

when the complex solution was allowed to pass at 1 cm^3 min^{-1}
until the said complex was first detected in the effluent,
relates to the foam structure. For Al- and A2-type open-cell
polyether foams (each loaded to about the same extent with TBP),
the breakthrough capacities were 12.1 and 16.7 g kg^{-1}, respectively [21], compared to 30 g kg^{-1} for an unspecified foam [17].

The separation of palladium from other metal ions has also
been investigated [17,21]. Thus, palladium was kept constant
(0.618 mg) and nickel varied by amounts up to 1000 mg in a
thiourea-perchlorate medium [21]. Sorption of the metal ions
was effected at a flow rate of 1 cm^3 min^{-1} and chromatography
at 1 to 4 cm^3 min^{-1}. These flow rates are considerably higher
than for granular supports and additional pressure was not
needed [17]. The palladium, unlike nickel, forms a complex
with thiourea and was completely taken up by the TBP foam even
from solutions with the highest nickel content. Nickel, but
not palladium, was easily eluted with perchlorate-thiourea and
the palladium subsequently removed with water presaturated with
TBP. Saturating the eluents with TBP is a most convenient way of
maintaining the original loading on the foam support [17]. Both
the elution peaks were again sharp and symmetrical [17,21].

In another set of experiments, bismuth was easily separated
from palladium although both form thiourea complexes, but bismuth was first eluted with 0.5 M perchloric acid followed by
palladium using the usual TBP-saturated water [21]. On the
basis of this trial work scheme, nickel (1 mg), bismuth (1 mg),
and palladium (0.618 mg) were recovered with 100% efficiency
from their mixture [21]. Again, elution peaks were sharp and
symmetrical, and the TBP-loaded columns could be used repeatedly
(>30 elution cycles) without a drop in efficiency. Moreover,
columns loaded with TBP do not deteriorate even if unused for
20 days [21].

The high HETPs of such foams in columns are not conducive
to high-resolution chromatographic-type separations. Despite
this fact the use of suitable reagents, e.g., thiourea-perchlorate,
facilitates the complete separation of palladium on the microgram
scale [21].

Bismuth and nickel can also be separated after sorption by
TBP foam from thiourea-perchlorate solution. Nickel is then
readily eluted by the background solution followed by bismuth
with 0.5 M perchloric acid [39].

The maximum extraction of bismuth was reached after shaking
with foam (56.4% m/m TBP) for about 10 min but over 90% of this
amount was extracted within 1 min [40]. The distribution ratio
at an aqueous concentration of 100 ppm bismuth was 33 dm^3 kg^{-1}.
The elution patterns are again symmetrical (Fig. 3) and also
HETPs are high being 2.8 mm at an eluting agent flow rate of

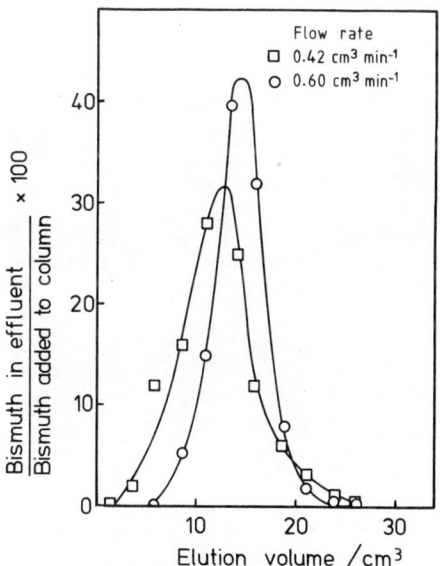

FIGURE 3   Elution behavior of bismuth(III) on TBP-loaded
polyether foam (46.8% m/m) at two different flow rates.
Eluent 0.5 M hydrochloric acid.

$0.42 \text{ cm}^3 \text{ min}^{-1}$ as calculated from the elution curve and 1.6 mm
from the breakthrough capacity curve (Fig. 4). The corresponding
HETPs at an elution flow rate of $0.6 \text{ cm}^3 \text{ min}^{-1}$ (Fig. 3) were
1.2 mm in each instance [40].

In chloride media it is possible to separate iron(III) from
cobalt(II), copper, and nickel on TBP-foams, but the latter three
cations cannot be separated from each other [22]. Nonetheless,
the separation of iron from cobalt is of some practical value
since the γ peaks of ^{58}Co and ^{59}Fe are too close for resolution
with a scintillation crystal. However, the carrier-free ^{59}Fe
isotope could be separated from the cobalt target material, as
adjudged by γ counting, after using TBP-loaded foam [22].

The distribution ratios for cobalt(II), copper, and iron in
hydrochloric acid up to 10 M have been measured for TBP-loaded
polyether foams and for classical partition using liquid

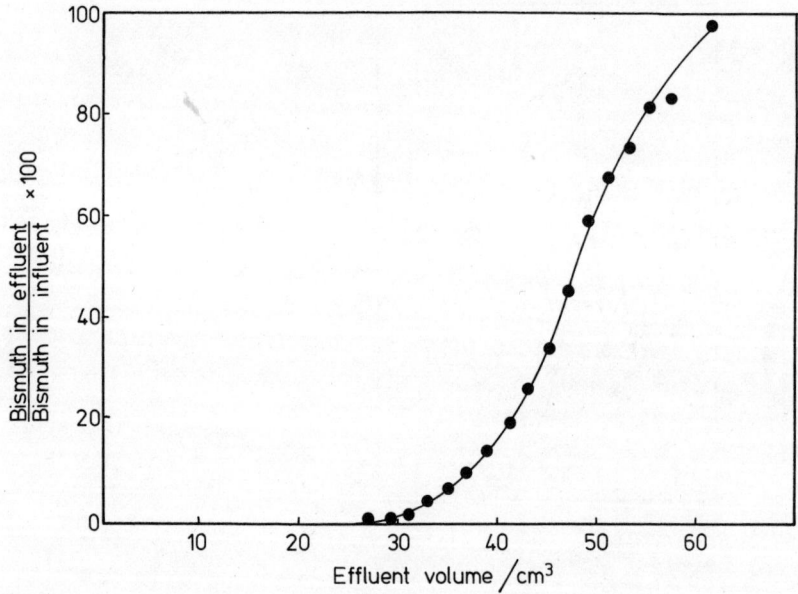

FIGURE 4   Breakthrough curve for bismuth(III) on TBP-loaded
polyether foam (46.8% m/m).

tri-n-butyl phosphate. The results are interesting from a
mechanistic standpoint since the distribution ratios and the
optimal acid concentrations found for each of the aqueous-
liquid TBP systems are not significantly different from the
corresponding TBP-foam batches [22]. However, the stability of
the foams at the high (10 M) hydrochloric acid levels is quite
remarkable.

Reverse-phase foam chromatography has also been practiced
to investigate the chemical enrichment of gold from dilute
solutions [25]. Thus, 95 to 97% of gold (0.1 mg) in a thiourea-
sodium perchlorate-perchloric acid medium (1 dm^3) was retained
on passing through a 5-g column of TBP foam at 50 to 60 cm^3 min^{-1}.
As in the case of the palladium-thiourea complex [21], the adsorption isotherm showed a good linear relationship over a relatively
wide concentration of gold. Washing with acetone, methyl ethyl
ketone, ammonia (2 M), or mineral acids (3 M) failed to completely elute the gold from the column, but quantitative recovery
was possible by dissolving the gold-foam material in hot concentrated nitric acid [25].

The gold-thiourea ratio of the extracted species calculated
by the slope method indicated the formula of the complex to be
$Au(H_2NCSNH_2)^+$. Furthermore, the distribution of this complex
was also determined at different concentrations of tri-n-butyl
phosphate, with toluene as diluent, when a log D-vs.-log [TBP]
plot indicated the probable formula of the extracted species to
be $[Au(H_2NCSNH_2)]ClO_4 \cdot 4TBP$.

Elements which can interfere with the analytical and industrial separation of gold include zinc, cobalt(II), nickel, copper,
iron(III), antimony(III), bismuth, palladium, and silver. These
can be categorized under the experimental conditions employed in
this work [25] as follows:

1. Zinc, cobalt(II), and nickel which do not form thiourea
    complexes

2. Copper, iron(II), and antimony(III) which form thiourea
    complexes which are not extracted or are only partially
    extracted with TBP-loaded foams
3. Palladium, bismuth, and silver whose stable complexes are
    extensively extracted by the TBP-loaded foam

Braun and Farag chose zinc, iron(III), and bismuth to
represent all these kinds of elements in their interference
studies [25]. The fast uptake of gold is hardly affected by
their presence. Thus, the half times of equilibrium adsorption
for gold alone and gold in the presence of zinc, iron(III), or
bismuth were found to be 0.4, 0.6, 1.0, and 0.9 min, respectively [25].

On short columns of TBP foams, trace amounts of gold(III)
could be separated from high levels of zinc, cobalt(II), nickel,
iron(III), antimony(III), copper, bismuth, and palladium by
retention from a thiourea-perchlorate medium. However, the
silver-thiourea complex could not be completely removed without
affecting the gold complex [25].

Uptake of the gold-thiourea complex by each of the unloaded
and TBP-loaded pairs of foam supports shown in Table 9 is essentially similar except that the half times of the equilibria are
shorter for the loaded supports [25]. Moreover, the recovery of
gold on the five different TBP-loaded foams showed a very small
range, namely, 90.0 to 93.5%, which again suggests liquid-liquid
partition involving tri-n-butyl phosphate as superimposed on the
now inert foam.

While 0.2 g of active carbon is about equivalent to 10 g
of loaded or unloaded polyether foams on a $t_{\frac{1}{2}}$ and percentage
gold uptake basis [20], the use of cheap, low-density foams
eliminates the troublesome, costly high solvent throughput
associated with such carbon systems. For industrial applications
unloaded open-cell polyether foams can be used instead of carbon
for the recovery of gold as a thiourea complex, but for laboratory
work the TBP-loaded foam is recommended because interference

TABLE 9 Recovery of Gold(III) from Perchloric Acid-Thiourea
Media with Six Different TBP-Loaded Polyurethane Foams, Voltalef
Powder, and Active Carbon

Type of foam[a]	TBP loading (% m/m)	Recovery of gold on sorbent medium (%)			
		Unloaded material		TBP-loaded material	
Polyether H.1 >95% open cell	66.6	93.2	(1.5)[b]	93.5	(1.0)[b]
Polyether H.2 >95% open cell	68.7	91.3	(1.3)	92.5	(0.5)
Polyether W8100 ∿100% open cell	65.4	89.0	(2.0)	91.5	(0.6)
Polyether W8300 ∿100% open cell	65.8	87.8	(1.5)	91.5	(0.6)
Polyether W8600 ∿100% open cell	68.4	85.2	(1.6)	90.0	(0.6)
Polyester PPI-80nFR ∿100% open cell	50.7	77.2	(1.8)	88.3	(0.4)
Voltalef powder	--	0		87.8	(0.8)
Active carbon	--	95.0	(0.6)	--	

[a]Cell dimensions of foams increase from type Hl to W8600.
[b]Parenthesized values are half lives (min) for adsorption of
gold-thiourea complex.
Source: Ref. 20.

effects are more easily controlled [20]. The utility of TBP as
a plasticizer [27-29,32] for chelate-based foams will be described
in Sec. 2.2.2.

    Apart from reversed-phase chromatographic art with tri-n-
butyl phosphate, only foams loaded with diethyl ether, methyl
ethyl ketone, 1-methyl ethyl ketone, or ethyl acetate have been
briefly examined for the extraction of trace amounts of gold
from aqueous solution [16].

2.2.2 Foam-Supported Chelate Extractants

Chelate extractants can be supported on foams with the aid of
plasticizers such as tri-n-butyl phosphate [27-29,32,41], α-di-
n-nonyl phthalate [26,27,36,39], di-n-dioctyl phthalate [27],
and dibutyl adipate [18]. Most chelates are, however, supported
directly on the foam matrix (Table 10). These include dimethylglyoxime [13,52], 1,2-ethanedithiol [15], dithizone (diphenyl-
thiocarbazone) [26-28,32,41,49,55], diethyldithiocarbamate (DDTC)
[14,28,29], benzoylacetone [54], 1-(2-pyridylazo)-2-naphthol
(PAN) [36,37,39,42,51,130], and 1-nitroso-2-naphthol [29].

Dithizone is a good complexing agent for mercury(II),
although typically of chelates, not completely specific:

$$Hg^{2+} + 2\ S{=}C\Big\langle\begin{array}{c}\overset{H}{N}-\overset{C_6H_5}{NH}\\[4pt]N{=}N\\C_6H_5\end{array}\ \rightleftharpoons\ \left[S{=}C\Big\langle\begin{array}{c}\overset{H}{N}-\overset{C_6H_5}{N}\\[4pt]N{=}N\\C_6H_2\end{array}\Big\rangle\right]_2 Hg + 2H^+ \qquad (18)$$

Nonetheless, some degree of this specificity is expected
to be conferred on the chelate loaded foams and which in turn
retain high-quality hydrodynamic character. These twin features
have facilitated some interesting separations and preconcentra-
tions in metal ion systems even when large quantities of other
ions are present. The classic, tedious liquid-liquid extraction
operations are, in effect, replaced by a solid-liquid system
based on both batch and column techniques.

Chow and Buksak [55] have shown that mercury(II) chloride
is best extracted by dithizone loaded foam at pH 5 to 10 compared
to pH 1.2 for methyl mercury(II) chloride and at low flow rates,
e.g., 0.3 cm min^{-1}. Recoveries of methyl mercury at levels from
0.002 to 500 ppb in control samples were >98.2%. Even in the
presence of 10 to 1000 ppm of aluminium, iron, zinc, copper,
magnesium, or calcium, the recoveries of both mercury species

TABLE 10 Selected Studies of Metal Ions on Chelate-Loaded Polyurethane Foam Systems

Source and type of foam	General pretreatment and foam loading technique	System	Analysis technique	Ref.
Open-pore polyurethane (ρ = 19.2 to 21.6 kg m^{-3})	Foam plugs 40 mm diameter by 53 mm long. Soxhlet with acetone for 6 hr. Air-dried. Soaked in benzoyl acetone dissolved in acetone, drained, and vacuum-dried.	Cd and Cu	Atomic absorption spectroscopy	[54]
Polyether based on polyol of mol. mass 4800 and MDI at OH/NCO = 1:0.3; Freon-water blown	Cubes ∿5 mm edge or cylinders 1.7 × 3-6 cm long. Washed in 1 M HCl; water; acetone. Dried at 80°C. Shaken with dimethylglyoxime for 1 hr. Dried between filter papers.	Ni	Spectrophotometry	[52]
Polyether foam (ρ = 20.9 kg m^{-3}) (Vert Foam Components, Manchester)	Cubes ∿5 mm edge or cylinders 15 × 15 mm long. Washed in 1 M HCl; water. Dried; washed with acetone, air-dried. Equilibrated with 5% m/v Na·DDTC in CCl$_4$ for 30 min. Excess reagents squeezed out between watch glasses. Na·DDTC loading was 0.14 mmol g^{-1} foam.	Sb(III) and Sb(V)	^{124}Sb	[14]
	As in Ref. 14 except air-dried foam shaken with 1% v/v of 1,2-ethanedithiol in benzene for 15 min. Extracted species is probably Sb(CH$_2$CH$_2$S)Cl.	Sb(III) and Sb(V)	^{124}Sb	[15]
Polyester foam DiSPo (Canlab)	Cylinders 4 × 3.5 cm long. Soxhlet with acetone for 24 hr. Acetone squeezed out and air-dried. Contacted with PAN-CHCl$_3$ for 24 hr. Orange foam air-dried and washed for 24 hr in water. PAN loading was ∿54% m/m.	Cu; Zn; Hg(II)	Atomic absorption spectroscopy; spectrophotometry	[57]
		Cd	Atomic absorption spectroscopy	[130]
Open-cell polyether foam (N. Hungarian Works, Sajóbábony)	Cubes 5 mm edge washed in 1 M HCl; water; acetone. Dried at 80°C. Stirred with silver dithizonate/TBP for ∿1 hr.	Ag	^{111}Ag	[32]
	As in Ref. 32 except dried-foam stirred with TBP/1-nitroso-2-naphthol.	Co(II)	^{58}Co	[29]
Polyester foam DiSPo (Canlab)	Soaked in 10 mg % dithizone/acetone for 1 hr. Drained and dried overnight in vacuum dessicator.	Hg(II) CH$_3$HgCl	^{203}Hg	[55]

61

at 0.002 ppb exceeded 96%. The possible effect of these six
cations was studied because of their common occurrence in natural waters and sewage. The extraction of each mercury species
from domestic water, river water, and various sewages was considered acceptable, especially for methyl mercury [55]. Each
mercury species is firmly bound to the foam and cannot be removed
with just water although up to 50% of the methyl compound is
eluted with 3 M hydrochloric acid. However, both mercury species
are quantitatively removed with acetone as well as an unspecified
quantity of the loaded chelate [55].

    The collection of mercury(II) and silver on TBP-plasticized
and nonplasticized zinc dithizonate loaded foams has also been
examined [28]. The overall capacity for mercury(II) nitrate was
22.3 meq kg^{-1} and leaching was offset by column washing with
TBP-saturated solutions [28]. The collection of traces of
mercury(II) or silver on the plasticized foam was generally
better than on the unplasticized foams. The initial collection
rate for silver decreased in the order α-Di-n-nonyl phthalate $\sim$
tri-n-butyl phosphate > di-n-dioctyl phthalate compared with
tri-n-butyl phosphate > α-di-n-nonyl phthalate > di-n-dioctyl
phthalate for mercury(II) but in all six cases the uptake was
finally about 100% for each cation after shaking for about
40 min [27,28]. The plasticizers were not leached from the
loaded columns even at flow rates of 50 to 60 cm^3 min^{-1}, and
the red mercury dithizonate foam was unaffected by air or light
[27]. Preconcentration of silver (1 ppb) and mercury (0.72 ppb)
was achieved on these loaded columns at 8 to 12 cm^3 (cm^{-2} min^{-1})
and with respective quantitative recovery using a 0.5 M sodium
thiosulphate [27,28].

Cadmium and lead can be recovered at pH $\sim$ 9 by dithizone foams (90% and 104%, respectively) at flow rates of 5 cm^3 min^{-1} and in turn elution was achieved with 4 M nitric acid [49].

Silver dithizonate foam plasticized with tri-n-butyl phosphate has been employed for the isotope-exchange separation of radiosilver:

$$AgHDZ(foam) + {}^{111}Ag^+(aq) \rightleftharpoons {}^{111}AgHDZ(foam) + Ag^+(aq) \qquad (19)$$

The average silver exchanged on the foam at flow rates of 3 cm^3 (cm^{-2} min^{-1}) and pH < 2 was >97.7%. In similar fashion a high quantitative exchange separation (>97.9%) has been achieved [32] on TBP-plasticized iodine-loaded foam at 3 to 4 cm^3 (cm^{-2} min^{-1}):

$$I_2(foam) + 2{}^{131}I^-(aq) \rightleftharpoons {}^{131}I_2(foam) + 2I^-(aq) \qquad (20)$$

Diethyldithiocarbamate (DDTC) [14,28,29] and 1,2-ethanedithiol [15] are two other sulfur-containing chelate ligands which have been studied in loaded foams. Thus, mercury(II) can be completely recovered from aqueous samples (0.72 ppb) on shaking with TBP-plasticized Zn•DDTC-loaded foams, over 97% being recovered after $\sim$10 min [28]. Complete and fast collection of cobalt(II) from solutions containing 1 ppb to 1 ppm by TBP-plasticized diethyl ammonium•DDTC foams at flow rates of 5 to 6 cm^3 (cm^{-2} min^{-1}) has also been reported. As with silver and mercury(II) on zinc dithizonate foams [27,28], the collection rates were faster using TBP-plasticized diethyl ammonium•DDTC-loaded foams [29]. Braun and Farag considered the sodium form of DDTC to be unsuitable owing to its low solubility in organic solvents [29].

However, batch and column experiments with Na•DDTC-loaded polyether foams established that antimony(III) could be separated from antimony(V) present in four types of waters after adjustment to pH 9.5 and when only antimony(III) is retained as Sb•$(DDTC)_3$. About 97% of the antimony(III) can be eluted with acetone at

1 cm^3 min^{-1}, and after drying the said foam, columns are ready
for reuse, but leaching of the DDTC chelate is not mentioned
[14].

The collection of antimony on polyether foams loaded with
1,2-ethanedithiol is independent of oxidation state and the type
of water involved, the concentration factor being 5000. Again,
the antimony can be removed with acetone at 1 cm^3 min^{-1}, but
unfortunately, the chelate is also leached [15].

The pH-dependent extraction curves for cadmium and copper
on benzoylacetone-loaded foam resembled those for the corresponding solvent extraction system. The extraction of copper
at pH 8.0 and cobalt(II) at pH 10.53 and at flow rates of <6
cm^3 min^{-1} was greater than 97.2% and 99.2%, respectively [54].

The optimum pH for extracting nickel using polyether foam
loaded with dimethylglyoxime is between 8 and 10 and is unaffected
by temperature in the 20 to 60°C range [52]. The adsorption is
quantitative down to levels of ∿0.5 ppm which is the limit of
classic gravimetric assay. As might be expected the highly
selective nature of this DMG chelate precludes any serious
intereference by iron(II), iron(III), zinc, calcium, and copper,
except for cobalt(II). Analysis of the nickel-foam systems
showed the average mole ratio of Ni/DMG in the foam to be 1:2.6
compared to the usual 1:2 ratio [52]. Nickel could be removed
from the foam using 1 M hydrochloric acid/ethanol (1:1 v/v).

The recovery of cobalt(II) at concentrations between 1 and
1000 μg on TBP-plasticized 1-nitroso-2-naphthol polyether foam
was quantitative and also faster than on nonplasticized foams
at flow rates of 5 to 6 cm^3 (cm^{-2} min^{-1}) and pH 6.8. Carrier-
free ^{58}Co was also successfully collected on the plasticized
foam column [29].

The chelate 1-(2-pyridylazo)-2-naphthol(PAN) is stable, of
low aqueous solubility between pH 2 and 11, and reacts rapidly
with many metal ions to form colored products. The prospects

of plasticized [36,39] and nonplasticized [37,42,51,130] foams
at typical loadings of 0.55% m/m and 0.35% m/m PAN [36,39] on
polyethers and polyesters have been extensively investigated,
e.g., the effects of 23 anions on the extraction of cobalt(II),
iron(III), and manganese(II) with plasticized foam. In general,
citrate, cyanide, and periodate mask each metal ion, whereas
borate, fluoride, phosphate, and tartrate mask iron(III) more
than cobalt(II) but have no effect on manganese(II) [36].

The extraction rate of cobalt(II) on PAN-polyester foam is
very high from aqueous thiocyanate media, but for iron(III) the
rate is highest from water. Anions which do not mask manganese(II)
do not affect the rate of extraction. Microgram quantities of
cobalt(II) are quantitatively collected on plasticized and nonplasticized polyester foam columns from potassium thiocyanate
solution at flow rates up to 10 cm^3 $(cm^{-2} min^{-1})$. For manga-
nese(II) complete recovery is only possible at <1 cm^3 $(cm^{-2} min^{-1})$,
but iron(III) is not quantitatively extracted even at flow rates
of 0.3 cm^3 $(cm^{-2} min^{-1})$. Cobalt(II) can be separated from
various amounts of manganese(II) at pH 4 to 5 since manganese(II)
moves with the solvent front. The cobalt(II) is then recovered
by elution with acetone [36].

The considerable effect of thiocyanate on the distribution
ratios for zinc, mercury(II), and indium(III) on unloaded and
PAN-loaded polyester foams is shown in Table 11.

The leaching of PAN from loaded foams depends on the pH,
the percent loading, the solution throughput, and the type of
foam. Thus, with solutions kept at pH 3 to 10 the loss of PAN
from a polyether foam loaded with 0.61% m/m PAN and 55% m/m
α-di-n-nonyl phthalate was negligible [39]. However, PAN is
completely leached from a nonplasticized polyester and one plasticized with α-di-n-nonyl phthalate [36]. The amount of PAN
leached from a nonplasticized polyether foam ranged from 7.5%
at 0.86% m/m PAN to 27.5% for the same foam at an initial

TABLE 11 Distribution Ratios for Zinc, Mercury(II), and
Indium(III) on Unloaded and PAN-Loaded[a] Polyester Foams

	Distribution ratio		
Metal ion	No KSCN	KSCN (0.2 M)	pH
Zn(II)	32 (2)[b]	22,627 (10,317)[b]	6 (2-3)[b]
Hg(II)	193 (277)	2,046 (15,052)	2-3 (6-8)
In(III)	4 (62)	6,273 (8,448)	2-3 (2-3)

[a]Loading = 1% m/m.
[b]Parenthesized values for unloaded foams.
Source: Ref. 42.

loading of 3.44% m/m PAN. The amount of PAN remaining after
leaching from foam at low initial loadings, e.g., 0.86% m/m,
is considered more than adequate for quantitative retention of
cobalt(II) traces [37].

Polyester foams loaded with PAN have been employed to
concentrate cobalt(II), zinc, cadmium, and mercury(II) from
large volumes of water [51,130]. Foams were squeezed auto-
matically in a tank apparatus operated by an 8-hr-life nickel-
cadmium battery when the orange PAN-only foam turned a reddish
purple color. The number of squeezes was also recorded and
related to the amount of cadmium, for example, extracted per
squeeze from four different standards (3.7 to 55.9 μg dm^{-3})
also containing species, namely, sodium, potassium, magnesium,
calcium, and chloride, commonly found in river waters. A plot
of the [μg Cd] extracted per squeeze vs. [Cd] in the tank then
enabled an estimation [130] of cadmium in unknown river waters,
possibly to within ± 10%.

Undoubtedly the most interesting application of chelate-
loaded foams concerns the semiquantitative, rapid, detection of
very low levels of metal ions and is based on their highly
colored complexes. This procedure has been tested in batch

and column modes using foam materials loaded with chromogenic
chelates which are called *Chromofoams* [26].

Thus, the semiquantitative assay of zinc was facilitated
by comparing the red color of its dithizone chelate, which
appears on an α-di-n-nonyl phthalate plasticized-dithizone cube
added to the sample (1 cm^3), with a standard color scale based
on 0.5, 1, 10, and 100 ppm zinc [26]. Thus at pH $\sim$ 6.2, 0.05 μg
of zinc can be detected compared with only 8 μg by the classic
dithizone spot test.

Dithizone foam columns have also been employed for the
semiquantitative assay of lead using standard leads ranging
from 50 to 500 ppb. The length of the red lead chelate zone
formed in the column is directly proportional to the lead in
the aqueous sample [26].

The same polyether foam loaded with thiocyanate-Amberlite
LA-1 ion-exchanger facilitated the determination of 1 μg of
cobalt(II) in the presence of up to 10 mg of more than forty
common ions. Copper(II) has been estimated from the dark green
chelate formed on rubeanic acid-loaded foams [26].

The sensitivity of these ingenious Chromofoams is better
than, or as good as, that realizable by normal spot tests [131].

The prospective use of other colored chelates, e.g., tin
(pink), mercury(II) (orange-red), and cadmium (orange) formed
on PAN-loaded polyester foams as Chromofoams has been suggested
[51].

The use of a styrene-divinylbenzene copolymer gel loaded
with a stabilized dithizone agency (DZ) is closely related to
the design of Chromofoams [132,133]. Thus, with a zinc-stabilized
0.01% dithizone gel column as little as 0.1 ppm mercury(II) could
be determined at a flow rate of 1 cm^3 min^{-1}. Samples containing
chloride (>100 ppm) present an interesting problem since the
mercury(II) is complexed as the chloromercury cation which reacts

in a 1:1 stoichiometric ratio [see reaction (18) for aqueous
mercury(II)]:

$$HgCl^+ + H_2DZ \rightleftharpoons HgCl \cdot HDZ + H^+ \tag{21}$$

Accordingly for saline waters the orange-colored zone is
roughly twice the length normally experienced with low chloride
samples on a comparable mass mercury basis and a separate high
chloride background calibration is necessary [132].

In similar fashion, when zinc samples (<1 ppm) were passed
through a green dithizone gel (prepared in an acetate buffered
thiosulfate medium) a sharp color change to the pink of the zinc
dithizonate complex was observed. The length of pink zone
enabled the determination of zinc down to 0.01 ppm [133] compared with 0.05 ppm using dithizone Chromofoam [26] at the same
pH level.

2.2.3 Foam-Supported Liquid-Ion-Exchangers

An anion-exchange foam containing about an equal mass of Amberlite LA-2 can be conveniently prepared by stirring a polyether
foam, D2, with about twice its mass of the exchanger, the excess
of which is readily removed by pressing the coated foam between
filter papers as described for solvent extractants (Sec. 2.2.1).
The rate of extraction of cadmium is intermediate between those
for coarse mesh (14-52) and fine mesh (52-100 and 100-200) solid
resins [38]. The effect of the degree of cross-linking on the
extraction is less significant than that of particle size, while
the equilibrium capacity of the foam and the resins is approximately the same (Table 12).

Apart from the anomalous behavior of iron(III) in hydrochloric acid between 4 M and 7 M, the extraction for the metal
ions shown in Table 13 parallel those for solid-resin exchangers
and solvent-extraction systems employing Amberlite LA-2. The
LA-2-coated foam extracts as efficiently as a 10% solution of

TABLE 12   Rate of Extraction of Cadmium (initially 0.5 mM) from
Hydrochloric Acid (2 M) by Five Solid Strong-Base Resins and by
a Polyether Foam Coated with Amberlite LA-2 (48% m/m)[a]

	Cadmium extracted (%)					
Time (min)	SRA71	SR56A	SR65	SRAK	Amberlyst	Coated foam
1	70.0	78.0	29.0	19.0	16.0	46.0
5	83.0	85.0	52.0	39.0	37.0	61.0
10	86.0	85.0	61.0	49.0	51.0	70.0
20	87.0	86.0	68.0	64.0	60.0	80.0
40	88.0	86.0	74.0	66.0	71.0	86.0
60	88.0	--	77.0	76.0	--	--
120	88.0	--	81.0	81.0	--	88.0
180	88.0	85.0	83.0	84.0	81.0	--
Mesh	100-200	52-100	14-52	14-52	45-55	
Cross-linking (%)	7-9	3-5	3-5	7-9		

[a]Amberlite LA-2 is N-lauryl(trialkylmethyl)amine.
Source: Ref. 38.

the same exchanger in 1,4-dimethylbenzene. Cobalt(II), copper,
lead, and nickel are not extracted by the foam to a significant
extent at any acidity. Quantitative separations are possible,
e.g., 100 μg of copper(II)-iron(III) and cadmium-zinc were
resolved on an 80 × 15 mm column containing 3 g of coated foam
at a flow rate of 1 cm^3 min^{-1}.

   Nickel and cobalt(II) in hydrochloric acid can be readily
separated even at relative ratios of $1:10^8$ by reversed-phase
chromatography on open-cell polyether foam coated with tri-n-
octylamine at 11.4 and 17.7% m/m loadings. Nickel and cobalt(II)
were eluted with 8 M and 1 M hydrochloric acid, respectively [24].

   The separation and concentration of radioiodine by isotope
exchange between $^{131}I^-$ and an open-cell polyether foam loaded

TABLE 13 Effect of Hydrochloric Acid on the Extraction of Metal
Ions (all initially 0.5 mM) by Amberlite LA-2-Coated Polyether
Foam

Hydrochloric acid (M)	Extraction (%)				
	Cd(II)	Fe(III)	Hg(II)	Sn(IV)	Zn(II)
0	0	17.6	64.4	95.1	13.1
0.3	64.7	2.4	96.4	53.5	10.6
0.7	71.0	8.5	99.1	--	25.6
1.0	75.0	17.1	97.5	67.6	29.2
2.0	88.0	58.1	94.5	77.2	54.4
3.0	72.0	80.5	92.1	83.3	55.9
4.0	65.4	86.5	89.0	87.8	48.7
5.0	56.5	85.8	83.8	86.9	36.1
6.0	35.5	83.9	73.1	84.3	24.0
7.0	18.5	85.2	56.9	83.4	16.6
8.0	7.0	93.8	43.5	78.7	12.1
9.0	1.6	87.0	30.9	72.6	8.2
10.0	0.0	98.1	--	67.4	7.0

Source: Ref. 38.

with iodine/tri-n-octylamine has been studied under static and
dynamic conditions [43]. Separations of about 90% efficiency
were achieved in about 10 min on such loaded columns (1 g) at
sample flow rates of 10 to 150 cm^3 min^{-1}.

In subsequent work [34,41,43-45] the high-quality exchanger
was replaced by a substantially cheaper technical tri-n-octyl-
amine-type Alamine 336 but without practical loss of separation
efficiency. Model experiments using ^{131}I paired with other
fission products, e.g, ^{89}Sr, [95]Zr, ^{137}Cs, [106]Ru, [140]Ba, or [147]Pm,
showed that except for traces of [95]Zr, only ^{131}I was retained on
the foam [43]. Radioiodine was similarly separated from fresh
and formaldehyde-preserved milks (1 dm^3) with about 8% efficiency

in 15 to 20 min [44]. These anion-exchange separations are
based on the fact that iodine, as I⁻, comprises over 95% of
the chemical species in the milk [44].

   Pulsating Column Separation

   The technique really depends on the special flexibility of
open-cell polyurethane foams, a character not matched in any
other kinds of inert supports. Of course, fast reactions at
the sample-foam interface are highly desirable. The technique
requires packing the foam loaded with Alamine 336 (or indeed
any other suitable reagent, e.g., chloranil [33] or TBP-zinc
dithizonate [41]) into a syringe (pulsed column). The process
is performed in two ways (Fig. 5). Typically, in the open
arrangement each fraction taken from the sample into the syringe
on the upstroke (A) is rejected on the downstroke (B) to waste,

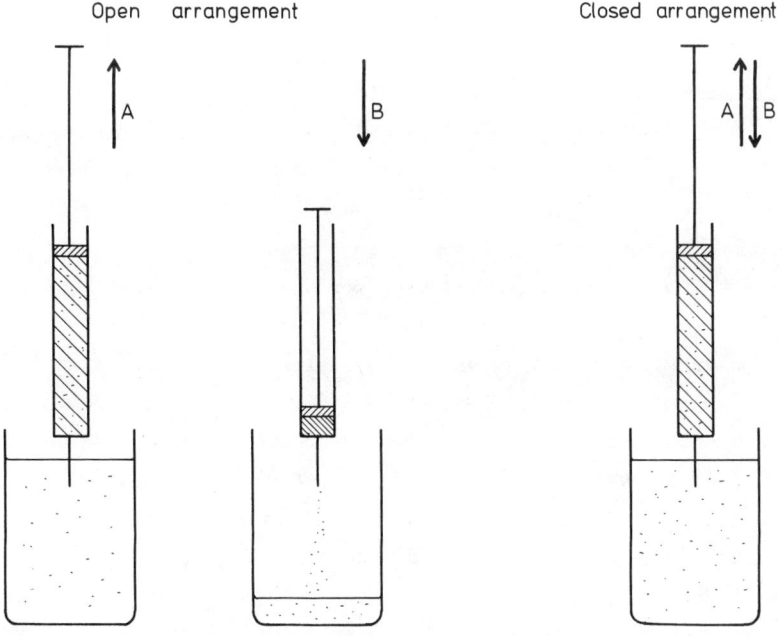

FIGURE 5 Principle of the open and closed pulsating-column
technique (A = charge; B = discharge). Reprinted with permission from T. Braun and S. Palágyi, *Analytical Chemistry*, 51:
1697 (1979). Copyright 1979 American Chemical Society.

and the process is repeated with further fresh fractions of
sample. For the closed arrangement the syringe content is
returned to the whole sample many times. In either event separation is performed by repeating the intake/discharge cycle and
each single cycle constitutes one pulse [41].

For the open system the efficiency of separation ($E_{n,op}$)
is related to the distribution coefficient (K) and number of
pulses (n) by

$$E_{n,op} = \frac{K}{n} \left[1 - \left(\frac{K}{1+K} \right)^n \right]$$

and similarly [41] for the closed system by

$$E_{n,Cl} = \frac{K}{n+K} \left[1 - \left(\frac{n-1}{n} \frac{K}{1+K} \right)^n \right]$$

These equations are useful in predicting the separation
efficiency for a selected number of pulses in a system of known
K. Thus, with ^{203}Hg, for example, the experimental value of 98%
obtained on TBP-plasticized zinc dithizonate foam, agreed with
the calculated $E_{n,Cl}$ values for chosen values of n = 5 to n = 25.
Efficiencies were about 93% for the nonplasticized foams [41].

The nature of the closed pulse operation lends itself to
automation of the plunger coupled to a counting device, and
which has been advantageously used for the routine assay of
radioiodine in natural waters [34]. After 500 pulses the separation efficiency was 97% while a 95% efficiency is achieved
more rapidly in just 200 pulses by simply raising the column
loading of tertiary amine by a factor of 2.5. The separation
efficiencies assessed by counting the activities of the aqueous
phase before and after pulsing [41] and also counting the foam
were in good agreement [41]. If necessary, over 95% of the
radioiodine can be eluted with acetone, dioxan, or toluene [45].
The ready-to-use iodine-tertiary amine foam is claimed to be
stable for longer than 10 days if stored in the dark [41].

The recent studies [51,130] on automatic collection of
heavy metal ions from large volumes of water is essentially
based on the pulsing technique.

2.2.4 Heterogeneous Ion-Exchange Foams

The application of columns filled with fine-mesh particles for
separation purposes requires forced-flow conditions to attain
reasonable flow rates. Heterogeneous ion-exchange foams, in
which finely ground commercial ion-exchange resins, e.g., Varion
KS, are incorporated into an open-cell polyether polyurethane
matrix, seem to resolve this flow problem. Foams containing up
to 40% m/m of such ion-exchangers match the mechanical properties of the original foams [92].

The use of polyurethane-Varion KS cation exchange foams
for fast separations has been reported [19]. Thus, the slope
of the breakthrough capacity curve for copper(II) is quite sharp
and sorption occurs in one fast step, $t_{\frac{1}{2}}$ being 0.6 min. The
selectivities of the heterogeneous Varion KS foam and original
KS cation-exchange beads are about the same. A complete separation of cadmium, calcium, iron(III), and zinc was effected on
the heterogeneous foam [19].

2.2.5 Foams Incorporating Inorganic Precipitates

The majority of materials used to load polyurethane foams has
been organic based but a few inorganic compounds have been
studied [30,31]. Thus, silver sulfide foam can be prepared by
loading the previously mentioned heterogeneous Varion KS foam
[92] with silver ion and then precipitating silver sulfide in
the foam matrix with sodium sulfide [30]. Columns of silver
sulfide foam retain radiosilver completely even at flow rates
of 20 cm^3 (cm^{-2} min^{-1}).

Foams containing finely divided copper(0) have also been
made by treating the Varion KS foam [92] with copper(II) sulfate

and reducing the copper(II) foam to copper (0) with sodium
dithionate [31]. Quantitative collection of radiosilver in
2 M nitric acid is again possible at flow rates of 10 to 12 cm^3
$(cm^{-2} min^{-1})$ on such foams [31].

2.2.6 Foams with Anchored Functional Groups

The preparation of polyurethane foams with specific functional
groups chemically bonded to its matrix represents an additional
advance in loaded foam design, particularly regarding operational lifetime since loss of the loaded agent by leaching
should, in principle, be eliminated.

Thus, a low-capacity, SH-polyurethane foam was prepared by
a rather complex treatment of a polyether foam with hydrogen
sulfide in a discharge tube. These foams which seem to be
stable for about 1 month have been utilized to concentrate
mercury(II) species from extremely dilute solutions [56]. The
collection efficiency for mercury(II) chloride between 0.0004
and 4 ppm was 99% compared with 80 to 50% for methyl mercury(II)
chloride over the same range. Mercury could be recovered by
Soxhlet extraction with 2 M hydrochloric acid [56].

Ionogenic groups have also been introduced into foams
indirectly [92]. Thus, anion exchange functionality can be
introduced into a styrene-polyurethane copolymer by classic
chloromethylation and amination [92]. A typical capacity of
such foams is 2.0 to 2.2 meq g^{-1}. An alternative approach is
based on grafting polyurethane foams with methacrylic acid
under the influence of fast electrons or ^{60}Co gamma photons
[92]. The resultant weak carboxylic ion-exchange foams had
capacities of 4 meq g^{-1}.

2.2.7 A Foam-Redox Column

Finely divided chloranil (tetrachlorohydroquinone) supported
on polyurethane foam columns, has been employed for reducing

cerium(IV), iron(III), and vanadium(V). Cerium(IV) can be
quantitatively reduced at room temperature at flow rates of
1 to 6 cm^3 (cm^{-2} min^{-1}), while reduction of the other species
is slower [23]. Packed columns could be used for over 30 reduction cycles without loss of efficiency and regeneration was
effected by washing with ascorbic acid [23]. Swollen foam
materials were more efficient than dry foams, and in each case
preheating the aqueous sample to about 80°C before pulsing
realized more effective reductions [33].

2.2.8 Foam Loaded with Di(2-ethylhexyl)phosphoric Acid

The extraction of metal ions by di(2-ethylhexyl)phosphoric acid
(DEHP) in organic solvents, e.g., benzene, is pH dependent. For
calcium in sodium nitrate solutions (μ = 0.5 M and 4.0 M) the
maximum extraction occurs at pH 5.5 and 5, respectively [134].

The maximum extraction of calcium by polyether foam loaded
with DEHP also occurred at pH 5.5. This extraction was conveniently monitored on a continuous basis using a PVC calcium
ion-selective electrode [135] coupled to a double-junction
reference electrode immersed in calcium chloride solution
(10^{-2} M) adjusted to pH 5.5 (Fig. 6). The final concentration
of calcium remaining at equilibrium after 2 hr closely matched
that determined by atomic absorption spectroscopy [60].

Any leaching from loaded foams discussed in Sec. 2.2
detracts from their utility. Therefore, the effect of several
different solutions on the loadings of DEHP foams has been
investigated [60]. Thus, DEHP-foams containing 3800 μg calcium
g^{-1} were squeezed once under 0.1 M hydrochloric acid and the
calcium content of the acid solution determined by atomic
absorption spectroscopy. This procedure was then repeated with
four further squeezing cycles with fresh 0.1 M acid, and the
calcium content of the foam plotted versus the number of squeezes
(Fig. 7).

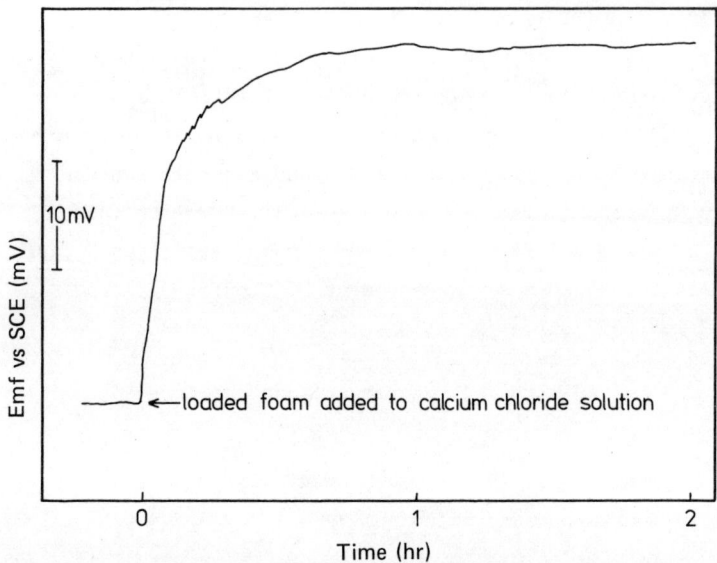

FIGURE 6 The uptake of calcium by DEHP-loaded foam as monitored
by a PVC calcium ion-selective electrode.

    The now partially calcium-depleted foam was reloaded from
10^{-2} M calcium chloride at pH 5.5, and the squeezing (leaching)
operation repeated (Fig. 7). Identical runs were undertaken
with 0.1 M solutions of acetic acid, dichloroacetic acid, and
EDTA and also with deionized water. Leaching was evident in
all cases. After each cycle of five squeezes, it was not possible to reload to the original calcium level of 3800 µg g^{-1} of
DEHP foam. After the sixth washing cycle with 0.1 M hydrochloric acid <50% of the original calcium could be reloaded.
This effect was clearly due to leaching of phosphate since
DEHP could be detected in the leaching solutions by thin-layer
chromatography.

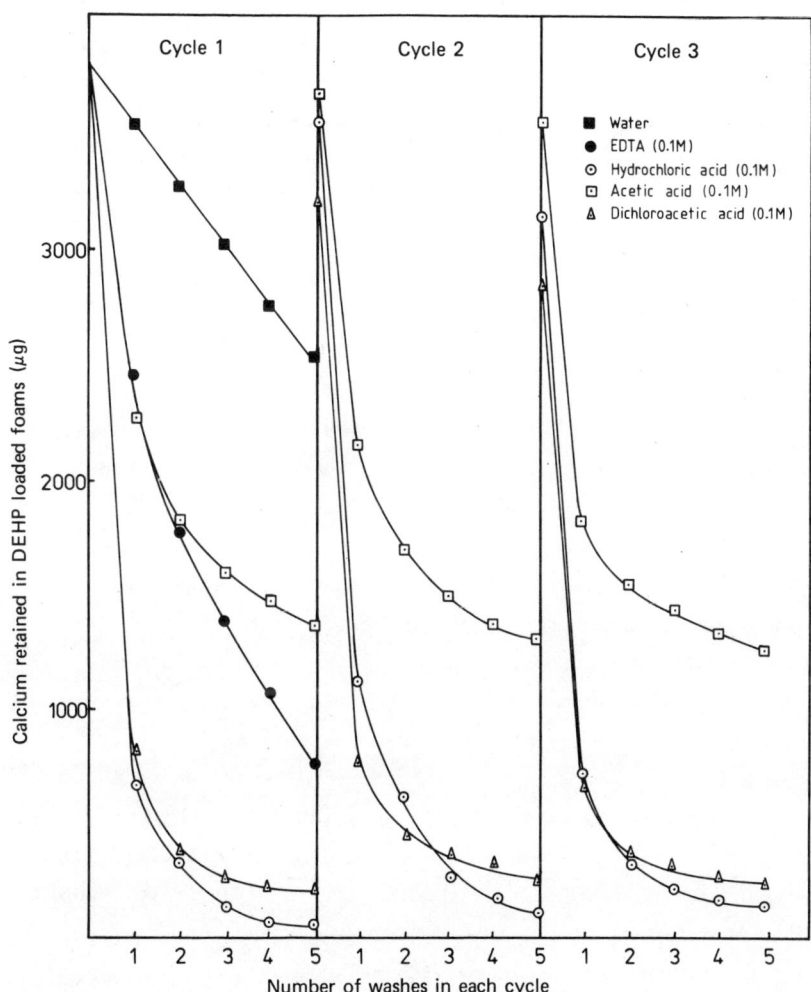

FIGURE 7 The loading/washing cycle of a DEHP-polyether foam
with five different reagents. Reprinted with permission from
T. Braun and S. Palagyi, *Analytical Chemistry,* 51:1697 (1979).
Copyright 1979 American Chemical Society.

2.3 SILICONE RUBBER-LOADED FOAMS

Open-pore silicone rubber foam has also been used as a chromatographic support [113,136] to a limited extent, but the stability in aqua regia for 24 hr is noteworthy, especially since
the catalyst, tin(II) octoate, remaining in the foam was eluted
with this acid [136]. This was a necessary pretreatment step
because tin(II) reduces palladium(II). Such tin-free foams
loaded with dimethylglyoxime permitted adsorption of palladium
(II) without serious interference from copper(II), iridium(IV),
platinum(IV), or rhodium(III). The adsorption was reduced to
some extent in the presence of 1000 ppm of chromium(III), lead,
silver, and especially iron(III) [136].

Baghai and Brown [113] also removed the tin(II) catalyst
from silicone rubber samples prepared as described by Gregoire
and Chow [136]. The presence of such metal ions in foam samples
is of special importance when trace metals sorbed on foams are
to be assessed by neutron activation analysis. Thus, the tin
content of a polyester foam sample was found to be about two
orders of magnitude lower than a polyether type of foam [42].

The tin-free foams loaded with tri-n-octylamine were pretreated with 6 M hydrochloric acid containing sodium chlorate
in order to provide some free chlorine and hence maintain the
metals concerned as rhodium(III) and iridium(IV). Without this
provision very little adsorption of iridium is possible [113].
In this way about 99% of the rhodium(III) remains in the sample
while 98.5 ± 0.9% of the iridium(IV) is retained on the foam.
This is quantitatively eluted with ethanol along with some
tri-n-octylamine which was also recoverable [113].

In both these investigations [113,136] the loaded silicone
rubber foams could be reused for up to 10 times.

3

The Concentration of Organic Compounds from Dilute Samples by Polyurethane Foams

This chapter essentially deals with the application of polyurethane foams for the concentration of various organic compounds, especially pesticides, in the atmosphere and waters.
Chromatographic applications are presented in Chap. 4.

3.1 PESTICIDES

Pesticides (insecticides) have been manufactured in relatively
large quantities, e.g., the total U.S. production of PCBs since
1957 of about 3.6×10^8 kg is estimated to be about one-half of
the total world production [137]. Aroclor 1242 and Aroclor 1248
are chlorinated biphenyls which together accounted for some 70%
of domestic PCB sales in the United States before 1970; by 1975
the pattern had changed to Aroclor 1016 (48%) and Aroclor 1242
(25%).

Organochloropesticides have conferred obvious benefits to
the management of pest problems throughout the world. Unfortunately, polychlorobiphenyls (PCBs) constitute a health hazard
to many animal species and to humans. Despite their low aqueous
solubility, e.g., 56 μg dm^{-3} for the PCB Aroclor 1254, concentration is effected via natural food chains and PCB levels in

some fish species in Lake Michigan have exceeded the Food and
Drug Administration guidelines of 5 μg g^{-1} for food in inter-
state commerce.

Some measure of the problem is indicated by the fact that
the weighted mean of 119 ng dm^{-3} for PCBs 'in 31 rainwater samples
taken on Lake Michigan in 1977 equate to an annual deposition of
5000 kg to the lake from wet precipitation [77]. On the other
hand, input from lakeside industrial plants employing PCBs for
heat transfer purposes, die-casting operations, and hydraulic
fluids has declined dramatically, presumably since their open
sale for such purposes stopped in 1971.

The use of the popular pesticide DDT presents similar environmental problems, e.g., its very high level of about 90 ppm
found in Koho salmon catches for Lake Michigan. Following the
ban of the use of DDT in the Lake Michigan basin in the late
1960s, the problem declined somewhat, e.g., the total DDT levels
in trout and bloaters fell from 19.2 to 10 μg g^{-1} and 10 to
2.1 μg g^{-1}, respectively, between 1970 and 1973.

The priority pollutant list established by the U.S. Environmental Protection Agency comprises 25 organochlorine compounds
which are considered environmentally hazardous if discharged
(indiscriminately) in industrial effluents (Table 1). The first
nine are multicompound mixtures, the simplest, chlordane being
the cis- and trans-isomers of 1,2,4,5,6,7,8,8-octachloro-2,3,3a,
4,7,7a-hexahydro-4,7-methaoindene. This is often used for struc-
tural pest control of termites and to a lesser extent for agricultural purposes.

Although single isomers of PCBs have been used in basic
environmental studies [69,70,76,81], PCBs are marketed as complex mixtures containing some of the 209 possible chloroisomers,
e.g., Aroclor 1242 is comprised of 54 identified isomers [138].
The isomeric nature of these products is rather interesting
since the most common method of analysis is based on electron

TABLE 1 Priority Pollutants List Published by the U.S. Environmental Protection Agency

	References[a]	
Pollutant	Atmosphere	Waters and Sediments
Aroclor 1016 ⎫	[69,76]	
Aroclor 1221 ⎪	[76,81]	
Aroclor 1232 ⎪		
Aroclor 1242 ⎬ PCBs	[76,80]	[49,62,77]
Aroclor 1248 ⎪	[70]	[49,68]
Aroclor 1254 ⎪	[70,76,80]	[49,68,72,73,74,77]
Aroclor 1260 ⎭		[49,61,68,72,77]
Chlordane	[71,79]	[71]
Toxaphene		
Aldrin	[80]	[49,62,66]
α-BHC		[49,61,62]
β-BHC		[49,62]
γ-BHC (lindane)	[79,80]	[49,61,62,66]
δ-BHC		
4,4'-DDD		[62,68]
4,4'-DDE	[80]	[49,62,66,68]
4,4'-DDT	[71,80]	[49,62,66,68,71]
Dieldrin	[79]	[49,62,66,68]
Endosulfan I		[49]
Endosulfan II		[49]
Endosulfan sulfate		
Endrin		[49,62,66]
Endrinaldehyde		
Heptachlor	[79]	[49,66]
Heptachlor epoxide		[66]

[a]Reference numbers refer to publications detailing analysis of atmosphere and waters following collection on, and elution from, polyurethane foams.

capture detection coupled with gas chromatography [80]. Thus, the sensitivity of the various isomers will vary by several orders of magnitude depending not only on the extent of chlorination but also on the ring positions of the chlorine atoms. Aroclor 1221 contains 51% of monochlorobiphenyl (Table 2) and three isomers are possible, e.g., 2-, 3-, and 4-chlorobiphenyl, respectively.

The remaining 18 pollutants listed in Table 1 are single-compound pesticides, e.g., γ-BHC or lindane, is 1a,2a,3a,4e,5e, 6e-hexachlorocyclohexane.

The determination of these ubiquitous materials in various waters treated with, or contaminated indirectly by, pesticides has naturally attracted considerable attention from environmentalists. Their content in the atmosphere, which is a major known route by which these substances are transported to waters, e.g., Lake Michigan [77], is also of interest. Before any analysis is undertaken some preliminary concentration step is

TABLE 2 Composition of Four Aroclors

Chlorobiphenyl	Composition of Aroclor type (% m/m)			
	1016	1221	1242	1254
$C_{12}H_{10}$	<0.1	11	0.1	<0.1
$C_{12}H_9Cl$	1	51	1	<0.1
$C_{12}H_8Cl_2$	20	32	16	0.5
$C_{12}H_7Cl_3$	57	4	49	1
$C_{12}H_6Cl_4$	21	2	25	21
$C_{12}H_5Cl_5$	1	<0.5	8	48
$C_{12}H_4Cl_6$	<0.1	a	1	23
$C_{12}H_3Cl_7$	a	a	<0.1	6
$C_{12}H_2Cl_8$	a	a	a	a

[a]Indicates <0.01%, none detected.
Source: Ref. 138.

essential since the levels involved are frequently extremely low,
e.g., 0.010 to 0.055 ng m^{-3} of 4,4'-DDT in the atmosphere over
the Sargasso Sea [71].

Thus, aquatic animals have been employed to monitor pesticide flow, e.g., freshwater mussels could effectively concentrate pesticides from water [139], although oysters have been
recommended [140]. Unfortunately, living organisms frequently
fail to survive in such contaminated zones and considerable
variation can arise in the performance between individual
animals.

One alternative approach to this collection-concentration
problem was based on a lengthy adsorption on charcoal from large
percolated volumes of sample water. While the adsorption is
efficient, the desorption step with large volumes of organic
solvent(s) can be difficult and inefficient, e.g., 25 to 30%.
Moreover, the catalytic nature of charcoal can effect significant chemical changes in the material originally sampled [141].
Excellent recoveries were reported using a reversed liquid-
liquid partition filter technique but was offset by low flow
rates (<65 cm^3 min^{-1}) and the need for vacuum backup [142].
Extraction of water samples with high-grade dichloromethane or
dichloromethane/hexane, followed by careful concentration of
the organic-rich layer, has facilitated efficient collections
of each of the 25 priority pollutants listed by EPA [143] (see
Table 1).

Prompted by Bowen's classic work [12,13], Gesser and co-
workers were the first to initiate the application of noncoated
polyester polyurethane plugs designed as culture tube stoppers
to this problem [63]. Since then foams have been used in pesticide studies on waters (rain, rivers, lakes, and oceans) [49,61-
64,66,68,71-74,77] and the atmosphere [69-71,76,79-81]. Apart
from occasional references to a particular type, e.g., Canlab
diSPo polyester foams [63,66,67], or their densities [63,69,70,

80], no detailed information is available on the nature of the
supports employed in the investigations. A number of studies
have been based on DC-200 coated foams [66,68,69,77,81].

3.1.1 Pesticides in Waters

As discussed in Chap. 2, for inorganic systems, both coated and
noncoated foams have been used in studies on pesticides present
in waters. The basic procedure adopted by Gesser and co-workers
[63] for water assay of PCBs involved several distinct stages:

1. Canlab DiSPo polyester plugs (density = 0.027 g cm^{-3} and
   2.2 cm diameter × 3.8 cm long) were cleaned by shaking with
   acetone-hexane (1:1 v/v).
2. Two such plugs were pushed down a glass tube (2 cm internal
   diameter), washed with ethanol to remove residual acetone-
   hexane, and finally washed with distilled water (200 cm^3)
   to displace ethanol.
3. The aqueous PCB sample was then passed down the column
   ($\sim$250 cm^3 min^{-1}) under gravity.
4. The foams were removed and eluted with acetone (20 cm^3) and
   then hexane (100 cm^3). The glass walls, which often adsorb
   considerable amounts of PCBs, were similarly washed.
5. The combined eluants were dried over anhydrous sodium
   sulfate and concentrated on a rotary evaporator, e.g., to
   10 cm^3, and a sample (5 mm^3) was analyzed in a suitable gas
   chromatography unit fitted with a ^{63}Ni electron capture
   detector (ECD).

This five-step protocol of collection-concentration,
elution, gas chromatography is rather typical of all other
pesticidal analysis. The solvents and foam quantities may be
different, e.g., up to 15 foam plugs may be used, elution may
be processed by Soxhlet extraction and mass spectrometry coupled
with gas chromatography (GC-MS) for identification of uncertain
GC peaks [76]. Thin-layer chromatography has occasionally been
employed to identify species [61,68]. Foams may also be attached
underneath a wooden float, an arrangement which keeps the foam
just under the surface of harbor waters [68]. For air samples
a pumping unit is essential, and the train of plugs is usually

protected at the intake end with a glass fiber prefilter to
occlude particulate matter.

It must be emphasized that in all these analyses the role
of the polyurethane foam is just to adsorb and concentrate the
organic materials--albeit to varying extents--from considerable
volumes of samples in a dynamic mode.

Gesser and co-workers found that 95% of PCBs in 1 dm^3 of
water was retained on the first foam plug and only a few percent
on the second backup plug [63]. Recoveries based on this tandem
plug column ranged from 91 ± 7% to 98 ± 6% and a complete analysis
was possible within <1 hr. It was also established that PCBs,
once concentrated on the plugs, were not washed out by water [63].

As already mentioned PCBs are multicomponent mixtures, and
this quantitation was based on a comparison of the peak heights
of 13 GC peaks with those of a standard Aroclor [63]. More
details of this technique and the alternative approach based on
derivitization of Aroclors are outlined in Sec. 3.1.3.

The extraction efficiency of DiSPo foams for Aroclor 1254
based on spiking experiments was about the same for distilled
water and filtered lake water but much lower for the unfiltered
lake water (suspended solids = 14 ppm) [74].

Musty and Nickless [49] reported recoveries of between 87
and 109% for 15 of the priority pollutants in Table 1 at flow
rates of 100 cm^3 min^{-1}, but only 40% for Aroclor 1260. At
higher flow rates [62], e.g., 250 cm^3 min^{-1}, with similar two-
pack foam plugs in glass columns, recoveries were much lower
(26 to 68%).

The foam material showed a very high capacity for these
pesticides, e.g., at least 100 mg of lindane could be passed
down two plugs (8 cm × 1.1 cm) without breakthrough. The fact
that concentrations of chloride ion up to 25 g dm^{-3} did not
significantly affect the performance of the foams [49] is
important for the analyses of saline waters.

The performances of noncoated polyurethane foams, Amberlite XAD-4 resin and Chromosorb W-n-undecane-Carbowax 4000 supports and a classical solvent extraction system have been evaluated for the analysis of four environmental waters [61]. Except for Aroclor 1260, the results were considered to be comparable (Table 3).

The adsorption of methylene blue from water has been employed as a criterion for the relative efficiency of extraction of pesticides from aqueous samples, the greater the amount of methylene blue adsorbed, the better the foam [62]. Of six different foams examined, the one capable of adsorbing about 250 mmol kg^{-1} of methylene blue in 24 hr has been recommended for this purpose [62].

Musty and Nickless also evaluated the recovery of several important pesticides on six different uncoated and coated foams [62]. In the majority of cases any one uncoated foam was more efficient for the recovery of a particular pesticide than when coated with DC-200 silicone (Table 4).

The suitability of a foam for extracting pesticides was based on the amount of methylene blue adsorbed, although the order of recovery of pesticides for the six uncoated foams A to F in Table 4 is not related to the quantity of methylene blue adsorbed, e.g., the efficiency of foam E is drastically different from foam D, and in four instances it is better than foam F. Moreover, no coated foam gave a better recovery than its uncoated form at the flow rates shown (Table 5) except at 10 cm^3 min^{-1} for endrin [62]. The fact that DC-200 coated Canlab diSPo polyester foam is superior to the uncoated foam for the extraction of nine pesticides is really a reflection on the foam type [66].

For coated foam studies the cleaned-up foam cylinders are dipped in the coating agency, e.g., 5% of DC-200 silicone grease in ethyl acetate [68], drained, and air-dried on absorbent paper.

TABLE 3 Levels of Three Pesticides in Various Waters Based on Four Different Concentration Methods

Concentration medium	Pesticide	Pesticide (ng dm^{-3}) at site listed			
		R. Severn at Sharpness[a] (2900 ppm Cl$^-$)	R. Severn at Broadoak[a] (3800 ppm Cl$^-$)	R. Leadon at Highleadon[b] (110 ppm Cl$^-$)	Long Ashton Research Station, Bristol University[c]
Amberlite XAD-4	Aroclor 1260	51	81	--	--
	α-BHC	--	3	5	53
	Lindane	--	30	21	91
Chromosorb W-n-undecane-Carbowax 4000	Aroclor 1260	64	13	--	--
	α-BHC	--	5	4	46
	Lindane	--	20	15	91
Solvent extraction hexane-acetone (41:59 v/v)	Aroclor 1260	135	675	--	--
	α-BHC	--	8	3	38
	Lindane	--	55	15	66
Porous polyurethane foam	Aroclor 1260	62	88	--	--
	α-BHC	--	3	4	38
	Lindane	--	40	16	79

[a]Fast-flowing tidal river, high in suspended sediment.
[b]Slow-flowing nontidal river, very low in suspended sediment.
[c]Rainwater run-off land sprayed with insecticide.
Source: Ref. 61.

TABLE 4 Recoveries of Nine Pesticides Using Six Foams of Varying Surface Areas

	Foam A		Foam B		Foam C		Foam D		Foam E		Foam F	
Pesticide	Uncoated	Coated[a]	Uncoated	Coated	Uncoated	Coated	Uncoated	Coated	Uncoated	Coated	Uncoated	Coated
α-BHC	101	83	61	46	64	61	65	72	57	52	70	65
β-BHC	101	91	62	48	63	77	77	82	63	66	66	66
Lindane	101	77	60	49	64	66	66	70	58	57	68	60
Aldrin	99	77	50	39	52	46	61	60	61	59	57	51
Dieldrin	106	88	71	58	85	69	94	96	70	72	78	76
Endrin	100	106	69	51	91	88	99	105	91	88	84	76
4,4'-DDE	106	81	70	58	96	79	103	103	93	97	83	76
4,4'-DDD	102	81	64	57	64	59	82	90	69	68	75	74
4,4'-DDT	114	88	47	50	68	61	64	73	69	68	59	44
Methylene blue adsorbed by uncoated foam after 24 hr (mmol kg^{-1})	260	--	96	--	70	--	156	--	28	--	116	--

Recovery (%) for foam listed at water flow rate of 10 cm^3 min^{-1}

[a]Each of the six foams coated with DC-200 grease.
Source: Ref. 62.

TABLE 5   Recoveries of Nine Pesticides Using Foam A Coated with Six Different Greases and at Two Different Flow Rates

	Recovery (%) on foam A coated with grease listed									
	QF-1		SE-30		DC-11		DC-200		Uncoated	
Pesticide	R1[a]	R2[b]	R1	R2	R1	R2	R1	R2	R1	R2
α-BHC	84	71	80	72	61	58	83	86	101	95
β-BHC	87	85	80	79	64	65	91	83	101	86
Lindane	83	76	81	74	56	52	77	77	101	91
Aldrin	71	57	58	59	65	62	77	67	99	73
Dieldrin	77	74	73	75	105	88	88	82	106	77
Endrin	95	76	87	67	73	69	106	91	100	94
4,4'-DDE	68	71	66	66	102	95	81	81	106	77
4,4'-DDD	73	70	68	73	89	92	81	80	102	89
4,4'-DDT	77	72	68	75	59	46	88	62	114	100

[a]Water flow rate = 10 cm^3 min^{-1}.
[b]Water flow rate = 30 cm^3 min^{-1}.
Source: Ref. 62.

Only hydrophobic agencies known to be clean were employed, that is, GC-grade greases. The efficiency of Canlab diSPo plugs coated with six different greases for removing a variety of pesticides from 4 dm^3 of distilled water (0.25 ppb) has been examined [66]. Extractions with all the different coated foams for light pesticides were generally higher than with the non-coated foam and for DC-200 was >92% in each case (Table 6).

Analysis of the first DC-200 plug in the train showed extraction efficiencies of 49 to 63% for the eight pesticides compared with 0 to 3%, 0 to 2%, and 0 to 1%, respectively, for the following three DC-200 plugs, while the glass column walls "extracted" 32 to 40% of the pesticides [66]. Recovery experiments using a river water individually spiked with 0.25 ppb of the pesticide ranged from 91 ± 7% for heptachlor to 100 ± 1% for lindane [66].

TABLE 6 Extraction of Pesticides from Distilled Water
by Polyester Foams Coated with Six Different Greases
(Flow rate ∿ 250 cm^3 min^{-1})

Pesticide recovered (%)	Foam coating						Noncoated foam
	SE-30	QF-1	DEGS	OV-25	OV-225	DC-200	
Lindane	95	100	100	97	91	97	55
Heptachlor	80	87	87	58	58	92	50
Heptachlorepoxide	95	95	89	77	76	100	68
Aldrin	71	80	73	47	45	92	45
Endrin	90	100	95	70	74	100	78
Dieldrin	60	73	51	48	44	100	73
4,4'-DDE	92	96	88	72	72	99	80
4,4'-DDT	50	69	34	45	45	98	84

Source: Ref. 66.

3.1.2 Pesticides in Air

The essential details of a typical collection unit for pesticides
in airborne samples is shown in Fig. 1. Uncoated, or coated,
foams (usually 2 to 4) of diameter slightly larger than the glass
or aluminum support tube for close fitting purposes are positioned behind a prefilter (Gelman glass type A) which precludes
airborne particulate matter. So for some fundamental studies
standard samples of single, or mixed, PCB isomers in hexane have
been placed on the first plug (1 cm long) followed by a train of
14 other clean, identical sized plugs. In this fashion, it is
possible to establish the distribution pattern and breakthrough of
any isomer along the train by drawing known volumes of pesticide-
free air through the prefilter. Pumping rates employed, e.g.,
0.2 to 1 m^3 min^{-1}, allow sampling of up to about 1500 m^3 of air
daily.

Some of the motor units in certain high-volume air samplers
carry capacitors located on the stator ring near the brushes

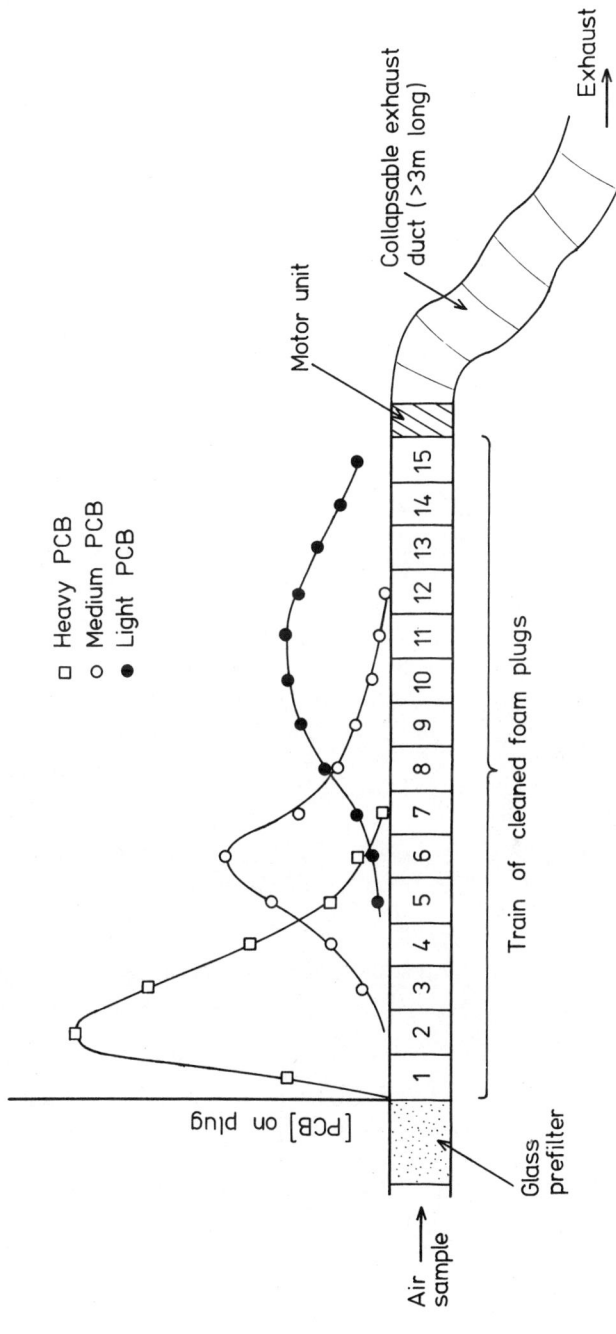

FIGURE 1 Essential components of air-sampler/collection unit and an idealized penetration profile along the plug train for three typical PCBs.

containing a polychloronaphthalene (PCN) and such compounds have
been detected ($\sim$1 μg dm^{-3}) on the foam plugs [144]. This implies
that at least some of the air is being recirculated and therefore
will give low assays for such air samples. Furthermore, PCN
(Halowax 1001) will coelute with any trapped pesticides and
cause serious interference in their analysis. Thus, only non-
PCN filled capacitors should be employed in these capacitor
devices designed to reduce sparking and line interference, but
are not really needed unless air collection is being carried
out near television or radio sets. Smoke bomb experiments also
demonstrated severe recirculation of the exhaust air to the
sampler intake.

The recirculation of sample is easily prevented by using
collapsable ducting, at least three meters long, on the high-
volume motor housing [76]. This in turn has eliminated many
extraneous GC peaks with low retention times which originated
in the sample exhaust. However, extraneous GC peaks can still
arise from contaminants in a laboratory area.

Bidleman and Olney [70] evaluated such a two-foam plug
sampler in relation to the collection of specific PCB isomers
at flow rates of 0.48 to 0.76 cm^3 min^{-1}. The following quantities were retained on the first plug: 3,4,2'-trichlorobiphenyl
(99%), 2,5,2',5-tetrachlorobiphenyl (96 to 98%), and 2,4,5,2',5'-
pentachlorobiphenyl (96 to 98%), each at the lowest sampling
rate. Bidleman and Olney concluded that polyurethane was a
suitable collection medium for airborne PCBs, although up to
4% of the penta isomer was retained on the GFA prefilter [70].

Penetration profiles based on the GC analysis of pesticides
in the individual foam plugs along a collection train is extremely
important to the design of units for real airborne analytical
samples. This is especially the case for Aroclors owing to the
volatility differences among the many isomers present in one
Aroclor. Thus, in a recent study a foam plug (7 cm diameter $\times$

1.0 cm thick) was spiked with about 40 µg of a mixture of known
isomers, or Aroclor 1016, and positioned at the head of a train
of 14 clean plugs of the same dimensions [69]. Air was then
drawn through the whole train fitted with a prefilter (in this
case comprised of the same foam material but of unspecified size)
at 0.15 to 0.81 m^3 min^{-1} to a total collection volume of 300 to
1600 m^3. Each plug was then analyzed by GC-ECD which clearly
established that each of the seven PCB isomers was eluted through
the foam in distinct bands and in order of decreasing volatility
or increasing molecular mass. Similarly, the order of elutions
for eight components in Aroclor 1016 paralleled their GC retention times, that is, the heavier components H and G with the
longest retention times are completely held on the first six
plugs, their highest levels being located on the second plug
(Table 7). All the isomer components, except for a small amount
of 3,3'-DCB, were effectively retained by the plug train at some
final point during the sampling period. The retention volumes
calculated even for the most volatile isomer, 3,3'-DCB, and for
2,4',5-TCB were 1250 m^3 and 2700 m^3, respectively [69].

Thus, the collection efficiency of foam relates strongly
to the volatility of any isomer, the total volume of air passed
and the length of the actual foam column. Despite the fact that
elution bands broaden as air flow rates are increased, there was
little difference in the thickness of foam required to retain a
given fraction of the PCB at the lowest and highest flow rates
tested that is, 0.15 to 0.81 m^3 min^{-1}. This is quite different
from the flow dependent patterns reported in water systems [61,
62]. Stratton et al. also observed that the collection efficiency of PCB vapors on polyurethane foams was flow independent
within the range of flow rates quoted for most common high-volume
sampling units [76]. Since isomeric species from mono- to hexa-
chlorobiphenyls are efficiently retained, it is reasonable to
expect that all others up to and including decachlorobiphenyl
will behave likewise [76].

TABLE 7 Penetration Profiles of Various Polychlorinated
Biphenyls along a Train of 15 Foam Plugs

Pesticide	Plug number with maximum PCB concentration	Number of plug beyond which PCB cannot be detected
3,3'-DCB	8	15, still contains PCB
2',3,4-TCB	3	9
2,4',5-TCB	3	10
2,2',5,5'-TCB	5	13
3,3',4,4'-TCB	3	9
2,2',4,5,5'-PCB	1	7
2,2',4,4',6,6'-HCB	1	5
A[a] Major GC peaks	10	15, still contains PCB
B in Aroclor 1016	6	12
C	5	11
D	5	9
E	3	8
F	2	8
G	2	6
H[b]	2	6

[a]GC peak with shortest retention time.
[b]GC peak with longest retention time.
Source: Ref. 69.

The collection efficiencies of the phosphorus-based
pesticides, diazinon[0,0-diethyl-0-(2-isopropyl-4-methyl-6-
pyrimidinyl)phosphorothioate] and malathion[0,0-dimethyl-S-(1,2-
dicarbethoxyethyl)phosphorodithionate] on polyether foams also
showed no significant change with sampling rate [80].

The thickness, or rather length, of foam needed to retain,
say, 90% of a given isomer presented in a sample of air can also
be calculated from the plot of fraction of total isomer retained
on the various plugs vs. plug number [69]. For 3,3'-DCP this

was plug 11 and thus a 11 cm length of foam would be needed for
an air sample collected at 0.5 m^3 min^{-1}. A few extra plugs in
a foam train would, of course, constitute a safety feature.

Plugs coated with silicone grease are reported to be more
effective than the corresponding noncoated foam plugs for scav-
enging pesticides from water samples [66], but this does not
seem to be the case for air samples comprising seven individual
PCB isomers on foams (density 0.022 g cm^{-3}) coated with SE-30 or
DC-200 silicones [69]. This is supported by a report that retention of ^{14}C-labeled 2,5,2',5'-tetrachlorobiphenyl on foams coated
with DC-200 silicone grease averaged 58% compared with 80% for
the corresponding noncoated foam [81]. Thus, the extra effort
and expense involved in coating foams and the problems associated
with eluting the concentrated pesticides from the coated foams
hardly seems a worthwhile exercise.

Plugs have been used repeatedly after cleanup for further
collections of pesticides but is a practice which is not recommended by Simon and Bidleman [69], who noticed that plugs used
for 12 successive runs appeared to be less efficient in retaining
PCBs.

3.1.3 Quantitation of Polychlorobiphenyls

In the case of single component pesticide samples the identity
and quantitation based on comparison with GC peak heights, or
areas, of the corresponding standard is relatively straight-
forward. For PCBs, however, this well-established procedure is
more difficult to apply owing to the very large number of peaks
associated with any one Aroclor, e.g., at least 16 GC peaks for
Aroclor 1016. The most volatile PCB species are likely to be
found in the atmosphere in greater amounts than the less volatile
components of any one Aroclor. Thus, except for four peaks, the
GC profile of an air sample collected on polyurethane foam in
the vicinity of a major industrial user of PCB closely resembled

that of an authentic Aroclor 1016 standard.   Two of the peaks
were tentatively ascribed to PCNs.   Examination of these PCB
patterns with GC-MS showed good correspondence between the real
Aroclor standard and the airborne sample for trichlorobiphenyl
and tetrachlorobiphenyl which of course constitute 78% of Aroclor
1016 (Table 2).   Furthermore, a comparison of the peak heights
of standards and samples clearly showed that the trichlorobiphenyl content is enriched in the sample relative to Aroclor
1016, whereas the tetrachlorobiphenyl is depleted in the sample
relative to Aroclor 1016.   This confirms the fact that the ratio
of trichlorobiphenyl to tetrachlorobiphenyl is significantly
greater in the air sample than the Aroclor 1016 standard.   Thus,
an air sample quantified on the basis of comparison with trichlorobiphenyl peaks in a standard Aroclor would be too high
and of course too low on the basis of a tetraphenyl comparison.

While an individual identifiable peak in the multicomponent
pattern of a PCB can be thus reasonably quantified as say
2,5,2',5'-tetrachlorobiphenyl, the environmentalist generally
wishes to identify and quantify the actual commercial Aroclor
product.   Some Aroclors are relatively easy to identify, e.g.,
Aroclor 1221 unlike the three other Aroclors in Table 2 contains
a high level of monochlorobiphenyls (51%), as well as biphenyl
(11%), and similarly Aroclor 1254 contains hardly any mono-, di-,
or trichlorobiphenyls and so can be distinguished from Aroclors
1016, 1221 and 1242 by examining the GC patterns.   However,
Aroclor 1016 cannot be unequivocally distinguished from Aroclor
1242.

Thus, an identity decision can often be made for an Aroclor
but its accurate quantitation is more difficult because of peak
pattern overlap and the effect of selective transport of trichlorobiphenyls compared with tetrachlorobiphenyls in the atmosphere.

Two methods have been developed for an accurate quantitation of PCBs present in the environment.   The actual method

employed is dictated by the appearance of the GC elution
patterns. The peaks in any pattern can be confirmed by comparison with the retention times of specific PCB isomers if
necessary.

The Pattern Matching Technique

The GC elution pattern is first subjectively compared with
a standard GC Aroclor pattern and quantitation then effected by
comparison with the peak heights and/or peak areas of several
peaks in that standard Aroclor. The assumption for both pattern
matching and for conversion of perchlorination results to the
equivalent Aroclor, say Aroclor 1242, is that Aroclor 1242 is
present alone, *and* in unmodified form, that is, no selective
enrichment due to volatility has occurred.

The Perchlorination Technique

The pattern matching technique is recommended when the GC
pattern of an airborne PCB matches a known Aroclor pattern.
Frequently this is not the case and the perchlorination procedure is adopted [76]. The technique recommended by Armour [145]
involves derivitization of all non-fully chlorinated PCB species
to a common, single isomeric species, namely, decachlorobiphenyl
(DCB), using antimony(V) chloride, e.g., for 2,5,2',5'-tetra-
chlorobiphenyl:

$$(1)$$

Provided 100% conversion is achieved then in one fell
swoop the GC pattern is greatly simplified in that the usual
16 plus peaks in an Aroclor 1016 chromatogram is replaced by
just one for DCB with a typical retention time of about 19 min
[76]. The detection limit is also enhanced since ECD is much
more sensitive to DCB than its lower chlorinated biphenyls.

Several problems arose [76] when attempts were made to evaluate PCBs using the Armour technique:

1. Perchlorination of mono-, di-, and trichlorobiphenyl isomers was not quantitative.
2. Although Armour effected about 100% conversion of Aroclors 1016, 1242, 1248, 1254, and 1260 at the microgram level this was not possible for < 1 μg quantities.
3. Hexane is also perchlorinated to a black mass (unidentified) which severely reduced DCB recovery.

The first two problems related to PCB losses during that step in the operation where the hexane-foam extract is evaporated to near dryness [76]. Hexane could be completely removed by three azeotropic evaporations from a hexane-chloroform mixture and the desired derivitization then proceeded successfully using the chloroform concentrate [76].

4. Biphenyl is also derivitized of course, eq. (1), albeit with low efficiency (43.3 ± 8.8%) [76]. Attempts to remove biphenyl before the derivitization step were unsuccessful and instead an allowance was made on the basis of a special GC run. Thus, biphenyl can be quantified by a gas chromatograph fitted with a flame-ionization detector. A correction based on the 43.3 factor for the perchlorination reaction is then made by subtracting that amount resulting from the conversion of biphenyl. Apart from Aroclor 1221 (11% biphenyl), this correction is probably unwarranted for the other Aroclors in Table 2.
5. Several commercial sources of antimony(V) chloride contained DCB which of course gives a high result anyway, while another contaminant, bromononachlorobiphenyl, interfered with the perchlorination of lesser substituted PCB species. However, the perchlorination reagent supplied by Cerac Pure, Milwaukee, was free from these compounds.

After the above modifications and the use of Cerac antimony(V) chloride perchlorination studies were undertaken [76] using a test mixture of 2-monochlorobiphenyl (25%): 2,2'-, 2,4-, and 2,4'-dichlorobiphenyl (55%) and 2,2',5-trichlorobiphenyl (20%) as well as with Aroclor 1016 and Aroclor 1254. The overall mean recovery for 18 runs was 100 ± 5%.

A Field Test for Polychlorobiphenyls

Having established the efficiency of the above procedure
for several known isomers and commercial Aroclor products, some
field trials were undertaken with foam trains attached to a
high-volume air sampler (see Fig. 1).

A landfill in New Bedford, Massachusetts, was known to
contain in excess of 22,000 kg of PCB containing wastes dumped
over many years until 1975 when it was closed.   Air sampling
was first undertaken on June 28 through June 30, 1977.   Thus,
after a 1-hr sampling period at 1.02 m^3 min^{-1}, only marginal
traces of PCB could be identified on the extract from the glass
prefilter as might be expected.   Apart from some 2,4'- and
4,4'-DDT peaks, the strong GC peaks from the extract of the
first polyurethane plug showed a close resemblance to those of
a full-blooded standard Aroclor 1242.   This sort of situation
of course is ideal for quantitative pattern matching and was so
calculated to be 0.89 μg m^{-3} of Aroclor 1242.   The second and
third plugs appeared to be devoid of quantifiable amounts of
this Aroclor product, or for that matter any product including
the DDTs.   These other pesticides must, therefore, have also
been adsorbed on the first plug but were not apparently evaluated
in the field trial [76].   No correction for biphenyl was necessary since it could not be even detected on the extract from the
first plug by GC-FID [76].

Extracts from the first plug samples collected over 1 hr
at the same site, that is, 61.2 m^3 of air were then perchlorinated.   The total amount of DCB found in the prefilter (173 ng)
and first plug (72,800 ng) was then converted [76] to equivalent
ppm Aroclor by

$$[72.97 \times \frac{\text{"molecular mass" of Aroclor 1242}}{\text{molecular mass of DCB}}] \div \begin{array}{l}\text{volume of}\\\text{air sample}\end{array}$$

that is,

$$\frac{72.97 \times 257.5/499}{61.2} = 0.63 \ \mu g \ m^{-3}$$

The discrepancy between the two values could be due to the mandatory assumption that Aroclor 1242 is present alone and in unmodified form [76] and once again serves to emphasize the problems associated with quantifying multiple-component DCB pesticides. Calculations for the perchlorination conversion to say equivalent Aroclor 1016 will fortuitously give exactly the same result since the *molecular mass* of the Aroclor is defined as the average whole number of chlorines calculated from the percentage chlorine substitution, namely, 257.5 in this case.

A subsequent sampling event at the same New Bedford site during January 1978, but now frozen and with light snow falling, indicated only 21 ng m^{-3}. This value was expressed in terms of Aroclors 1242 and 1016 because it was not possible unequivocally to distinguish between the two products [76].

It was concluded from these data that the landfill is a possible low level PCB emitter during winter months compared with the summer [76].

Pesticides have also been measured in the marine atmosphere and surface waters of the Bermuda-Sargasso Sea zone of the North Atlantic [71]. Air samples were pulled through a single 15-cm plug of polyurethane foam (20-cm diameter) carrying a Gelman A glass fiber (20 cm $\times$ 25 cm) at 0.4 to 0.8 m^3 min^{-1}. The GFA filter removed 98% of particulate matter with radius >0.015 μm. The quantitative problem is again evident in that the GC patterns of air samples matched two different standard Aroclors and PCB values in Table 8 were calculated as Aroclor 1242 and 1248. Calculations were also quoted on the basis of Aroclor 1254 since most other workers up to that time had quantified the PCBs in terms of the GC patterns matching with Aroclor 1254.

On the other hand GC patterns of PCBs from Sargasso waters closely resembled Aroclor 1254 or 1260 which contain mainly

TABLE 8 Concentrations of Some Pesticides in Land and Marine Air Samples

Collection dates (1973)	Location land/sea	Air volume (m³)	PCB[b]	Pesticide (ng m^{-3})	
				4,4'-DDT	Chlordane(cis + trans)
5/8	Providence, RI	76	94	0.09	0.25
2/12 to 2/13	Bermuda	1070	0.59 (0.39)[c]	0.013	< 0.005
2/13 to 2/14		1320	0.30 (0.22)[c]	0.012	< 0.005
2/29 to 3/9		732	0.52 (0.29)[c]	0.032	0.008
6/4 to 6/5	33°20'N, 65°14'W[a]	300	1.6	0.016	0.090
6/9	40°32'N, 70°20'W[a]	196	0.83	0.055	0.17

[a]Collected from R. V. Trident in Sargasso Sea while ship en route to Rhode Island.
[b]Calculated as Aroclor 1242 or 1248.
[c]Calculated as Aroclor 1254 [70].
Source: Ref. 71.

penta-, hexa-, and heptachlorobiphenyls, and it was concluded
that the more highly chlorinated biphenyls are selectively
transported to the ocean or that the lighter chlorinated PCBs
are degraded more readily in seawater [71]. It was also interesting that the levels of pesticides were higher in the surface
microlayer (150 μm) than at 30 cm below the surface, e.g.,
5.6 ng dm^{-3} versus 1.6 ng dm^{-3} for Aroclor 1260 and 0.7 ng dm^{-3}
versus <0.15 ng dm^{-3} for 4,4'-DDT. The concentration of cis- +
trans-chlordane was <1 ng dm^{-3} for all water samples [71].
Turner and Glotfelty have shown that polyester foams trap 98%
of the trans-isomer in air samples [79].

3.2 HERBICIDES

The drift of pesticides and herbicides during and after spraying
is a matter of concern. Thus, the measurement of the concentration profiles of herbicides in the air over treated agricultural
zones indicates the extent of their evaporation with time.
Turner and Glotfelty have designed an aluminium mast assembly
fitted with 12 holders for foam plugs (4.5 cm diameter × 5 cm
long) at various height intervals for field work. Considerable
variation in air flow resistance was observed--even between two
plugs cut from the same polyester foam sheet--when measuring
the pressure drop required to maintain a face velocity of 70 cm
sec^{-1} across individual plugs. Only plugs with flow resistances
of <12 cm sec^{-1} were used; these graded plugs showed ∿10% variation in flow rates [79]. Simon and Bidleman have also reported
air-flow resistance to vary by as much as 30% for plugs cut from
the same sheet of an unspecified polyurethane foam [69].

Trials at the highest flow rate, 5 m^3 hr^{-1}, showed that the
first of two such plugs could trap 98% or more of the herbicides
trifluralin (I) and dacthal (II),

I   II

even for the highest quantity tested, namely, 300 µg. Moreover,
the trapping efficiency was maintained even when herbicide-free
air was drawn through the exposed plug for an additional 18 hr.
This simple test shows how well a material is retained when just
air is drawn through a spiked plug. Vapor penetration declined
as the flow rate was lowered and at the 3.5-m^3 hr^{-1} sampling
rate used in subsequent field work recovery for the first plugs
exceeded 99%. Detection limits based on 5-m^3 air samples were
~ 0.1 ng m^3 for trifluralin and considerably less for dacthal
[79].

Trapping efficiencies were also studied with p-chloronitrobenzene (CNB), a nonherbicide compound with a vapor pressure
some three to four orders higher than I or II. The empirical
expression for trapping efficiency

$$\frac{\log P_2}{\log P_1} = \frac{V_2 F_1}{V_1 F_2}$$

which related CNB vapor penetration P to air flow rate F and
foam plug volume V (where V = area × length) also appeared
applicable to data for the herbicides. This simple equation
indicates that to simultaneously increase recovery from 50 to
95% and also double the air sampling rate, then the plug volume
must be increased over five times, e.g., by lengthening the
foam train. Although this equation is considered a useful
approximation, Turner and Glotfelty recommended experimental
testing for every new set of conditions [79].

Field trials were then conducted with the mast assembly. For this purpose a homogeneous mixture of I and II plus trans-chlordane and heptachlor was uniformly sprayed on the soil of an experimental site in Beltsville, Maryland, using a tractor. After 1 hr, samples were taken simultaneously at 12 different heights above the treated soil and at air flows of 3.5 $m^3 h^{-1}$ for the next 36 hr. Near the ground the trifluralin level of $\sim$106 $\mu g\ m^{-3}$ measured after the first hour had fallen to $\sim$30 μg m^{-3} after 24 hr compared with a fall from $\sim$18 to $\sim$6 $\mu g\ m^{-3}$ over the same time interval at a height of 200 cm above ground. The height-time profiles for the other compounds were very similar and only differed in concentration [79].

The only difficulty encountered in these valuable field trials was the preparation of enough uncoated plugs, and of course, this problem would be compounded if coated plugs had been employed.

A most interesting sideline investigation was also very briefly mentioned, namely, the suitability of the plugs for sampling the gypsy moth pheromone, disparlure [(Z)-7,8-epoxy-2-methyloctadecone], and also benzo(a)pyrene. Turner and Glotfelty believe that polyurethane foams will prove to be a most versatile trapping medium and that similar unusual air sampling applications will be forthcoming [79].

3.3 PHTHALATE ESTERS

Phthalate esters are used extensively as plasticizers in many polymer products and, although not particularly toxic, pose a threat to the environment. Their extraction by polyurethane foams from water [65] and air [78] has been reported.

In basic recovery studies water (100 cm^3) was spiked with a single ester (0.1 μg) and run on a five-plug foam column at 10 $cm^3\ min^{-1}$. Each foam was then extracted with acetone (2 $\times$ 2 cm^3) and then hexane (2 $\times$ 3 cm^3) and 4-μl samples

TABLE 9   Recovery of Nine Different Phthalate Esters on Two
Different Polyurethane Foams

	Recovery of ester (%)		
	Foam A (ρ = 0.030)		Foam B (ρ = 0.019-0.022)
Phthalate ester	Coated[a]	Uncoated	Uncoated
Dimethyl phthalate	13	20 ± 5	8 ± 1
Diethyl phthalate	70	84 ± 8	80
Di-n-butyl phthalate	98	102 ± 4	101 ± 1
Diisobutyl phthalate		93 ± 2	99 ± 2
Dipentyl phthalate	49	85	80
Di-n-hexyl phthalate		34 ± 2	33 ± 2
Di-n-heptyl phthalate		23 ± 2	
Di-octyl phthalate	9		
Butylbenzyl phthalate	74	70 ± 3	86 ± 3

[a]Five plugs each coated with 5 mg of DC-200 silicone oil.
Source: Ref. 65.

submitted for analysis by GC-ECD (Table 9). Recoveries of
individual phthalates varied considerably and except for
butylbenzylphthalate foam A coated with silicone oil conferred
no special benefit on their recovery.

The atmosphere over urban Osaka, Japan, has been sampled
by pulling air at 28 dm^3 min^{-1} through two polyurethane foams
(2.3 cm diameter × 5 cm long; ρ = 0.024) for 48 hr followed by
Soxhlet extraction with light petroleum-hexane (5% v/v). The
minimum detectable concentrations for dibutyl, diheptyl, and
diethylhexyl phthalate by GC-ECD was 0.0041 to 0.011, 0.001,
and 0.0019 to 0.0036 μg m^{-3}, respectively, when 80 m^3 of air
was sampled at the EPA center. Recovery runs showed that these
esters were completely trapped in the first plug, average recoveries being 99.1, 99.3, and 92.9%, respectively. The actual

levels of the three phthalate esters during February, 1976,
over Osaka were 0.025 to 0.060, 0.005 to 0.006, and 0.027 to
0.036 $\mu g\ m^{-3}$, respectively [78].

3.4 ALKYL BENZENE SULFONATE

Polyether foam has been used for the analysis of alkyl benzene
sulfonates, specifically lauryl benzene sulfonate (LBS), which
in small amounts, was selectively adsorbed from an aqueous
solution of $2.5 \times 10^{-5}M$ crystal violet (CV) adjusted to pH 2.5.
The color intensity of the adsorbed LBS.CV ion-association complex was proportional to the benzene sulfonate and could be
visually related to the initial sample concentration by a standard series method [112]. This is really a chromofoam technique
(see Sec. 2.2.2).

Alternatively the colored complex was eluted from the foam
with methanol (20 cm^3) and the LBS determined spectrophoto-
metrically at 588 nm. Satisfactory results were claimed using
white polyether foam (ρ = 0.024) but foams of polyester polyurethanes, poly(vinyl chloride), polystyrene, and polyethene
were unsuitable. The analysis of a river water by the two procedures is shown in Table 10. The detection limit of 0.5 ppm

TABLE 10 Determination of Laurylbenzene Sulfonate in
River Water[a]

Sample (cm^3)	LBS added (ppm)	LBS found (ppm)	
		Chromofoam	Spectrophotometry
20	--	5	5.7
40	--	5	5.5
20	5	10	11.0
20	10	15	13.8

[a]Small river in Ikeda City, Japan.
Source: Ref. 112.

using this static method could be lowered to 0.1 ppm using a
polyether foam column (0.8 cm diameter × 5 cm long) at sample
flow rates of 20 to 30 cm^3 min^{-1}. Methylene blue could be
employed in place of crystal violet at pH 4.6, but the detection
was less sensitive [112].

3.5 PHENOLS

Refinery streams such as fluid catalytic cracked heavy gasoline
thermal gasoline, and light thermal cracked gas oil may contain
appreciable quantities of valuable phenolics, that is, benzene-
based compounds like phenol, aminophenols, chlorophenols, nitro-
phenols, the cresols, the xylenols, as well as naphthalene-based
compounds, e.g., the naphthols. Such materials are troublesome
pollutants in an industrial stream and should not be passed on
to successive plant streams or indeed to the environment. In
any case they constitute useful chemical products and their
economic recovery is desirable. Various adsorbents have been
used for recovering phenolics from plant streams but tend to be
costly and are associated with the usual problems of low capacity,
regeneration and short operational lifetimes of the adsorbent.

Schlicht and McCoy have therefore evaluated the adsorption
of phenolics from n-heptane (75 cm^3) after shaking with polyether foams [86]. The uptake of phenolics was monitored by
comparing the infrared band at 1600 cm^{-1} of the original solution with the supernatent liquid at the end of the shaking
period. In this batch method the foams with the lower nitrogen
contents appeared to be the most effective for adsorbing phenol
from n-heptane (Table 11). Bowen also reported that the distribution coefficients for phenol on five different polyether
foams to vary from 46 to 410 for an aqueous system [13].

Considerable recovery of phenol was also achieved by passing
the samples down an adsorbent column. The effluent was periodic-
ally monitored for phenol, and the stage at which the concentration

TABLE 11 Adsorption of Phenolics by Polyether Foams and Three Other Supports

Adsorbent (3 g)		Phenolic component	Uptake of component (%)
Polyether polyurethane foam	3.1% N		85
	5.6% N	Phenol	60
	5.1% N		65
	5.6% N	Mixed cresols	60
	6.8% N		50
Activated vegetable			45
Attapulgite clay		Phenol	None
Porcelalumina			None

Source: Ref. 86.

of phenol rapidly increased was defined as *breakthrough*. This simple procedure provides a comparative indication of the capability of an adsorbent for cleaning up various phenolics from a particular solvent (Table 12).

Acetone was used in both static and dynamic modes to elute adsorbed phenolics from the various adsorbents which, after air drying, were reusable for many subsequent runs [86].

This basic research established the following points regarding phenolic adsorption on polyether foams:

1. Adsorption efficiencies of up to about 90% are rapidly achieved for phenolics and in amounts about equal to the mass of foam itself. This contrasts with other adsorbents which are much less effective.
2. Regeneration is quickly effected with acetone.
3. Foams can be recycled many times through this simple adsorption-regeneration process.

TABLE 12 Breakthrough Volumes for Phenol on Nopea Polyether Foam and Attapulgite Clay

Phenol sample	Adsorbent	Flow rate (cm³ hr⁻¹)	Breakthrough volume (cm³)	Phenol uptake (g)
Phenol-n-heptane, 2.85% m/m	Ether foam (60 g)	300	2600	52 (0.86)[a]
	Ether foam (60 g) after third acetone recycle	600	2400	48 (0.80)
	Attapulgite clay (535 g)	500	1200	24 (0.045)
	Clay (535 g) after second acetone recycle	500	1050	20.5 (0.038)
Phenol-methylcyclohexane, 2.5% m/m	Ether foam (60 g) after an unspecified number of acetone recycled	600	2700	54 (0.90)

[a]Parenthesized values represent phenol uptake per gram of absorbent.
Source: Ref. 86.

3.6   NICOTINE

The irritating action of cigarette smoke is partly ascribed to
its content of tars, nicotine, aldehydes, hydrogen cyanide, and
sulfides.   Much effort has been expended in reducing, and ideally
eliminating, such undesirable constituents while at the same
time maintaining the taste (aroma) quality expected by the
smoker.   Jefferson and Salyer [2] have tested an open-pore
polyurethane foam prepared from the polyol LA-475 and crude
polyisocyanate MDI as a cigarette filter.   Smoke from various
(unspecified) brands of cigarettes containing tar and nicotine
was pulled slowly through each of five samples of the foam cut
to the same size as the cellulose filters commonly used as brand
filters (1.4 cm long × 0.7 cm diameter).   The same volume of
smoke was also drawn through each of five cellulose plugs removed
from the brand cigarettes.   The average gain in mass for the
polyurethane foam filters was 45% compared with 33% for brand
cellulose filters.

     A more extensive investigation has concerned the replacement
of the normal filters in ten brands of cigarettes by polyurethane
foam filters.   This foam was synthesized from a polyol (formed by
condensing sorbitol with propene oxide) and toluene diisocyanate
admixed with sodium tungstate as catalyst, ethyl silicate as chain
extender and trichlorofluoromethane as foaming agent.   The resulting rigid, granulated mass was then treated with hot water to
hydrolyze any residual isocyanate and ethyl silicate, both of
which impart an irritating taste to the tobacco smoke.   After
drying at 60 to 105°C, the material (150 mg) was rolled into
filter plugs and attached to the various brands stripped of
their original filter tips [87].

     A portion of the nicotine and particulate matter was removed
from the cigarette smokes using these attached, modified foam
filters (Table 13).   In addition, the filters reduced the moisture content and the temperature of the smoke streams.

TABLE 13 Uptake (mg) of Nicotine and Particulate Matter from
Cigarette Smoke by Polyurethane Foam Filters

Cigarette (filter brand)	Nicotine	Total particulate matter
Carlton	0.20	3.2
True	0.36	6.6
Kent	0.35	7.9
Lark	0.34	9.6
Winston	0.50	9.9
Viceroy	0.39	6.7
Lucky Strike	0.53	10.3
Marlboro	0.41	7.9
Chesterfield	0.53	10.7
Pall Mall	0.55	10.9

Source: Ref. 87.

Other additives such as aluminium hydroxide or magnesium
trisilicate can also be incorporated into foams either at the
time of preparation, or subsequently by intimate physical admix-
ture, with advantage in that smokers experienced a more acceptable (milder) taste [87].

3.7 OILS

Today oil spillages to bodies of water from tankers, storage
vessels, and pipelines are commonplace, and various techniques
have been devised for their cleanup. In this context, the
observation [2] that a test water layered with 10 cm^3 of SAE-10
motor oil could be collected in about 2 min by a floating block
of porous polyurethane foam ($\sim$5.8 cm $\times$ 3.5 cm $\times$ 1 cm) is indica-
tive of an important environmental application for foams which
are relatively cheap and manufactured annually in multimillion-
kilogram quantities.

Will and Grutsch have designed a motor driven inclined belt
of polyurethane foam for removing oil from contaminated waters

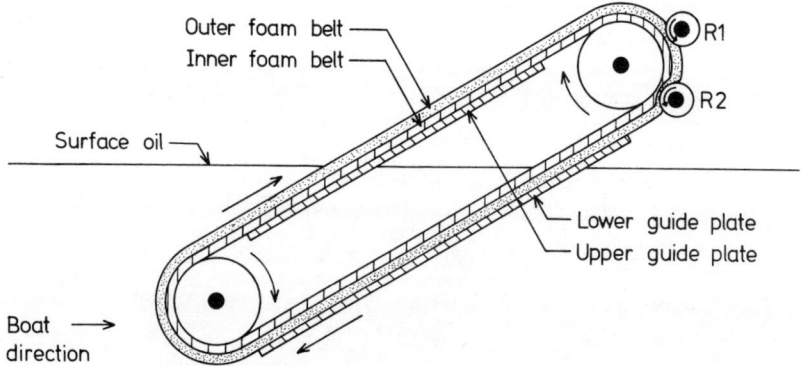

FIGURE 2 Moving foam belt unit for recovering oil from
contaminated water. (From Ref. 88.)

(Fig. 2). This unit with nylon-reinforced foam belts is particularly adaptable to handle very thin oil slicks, e.g, 25
barrels of oil per square mile [88].

The outer foam layer has a larger porosity than the
abutting inner layer, the whole being supported by underlying
guideplates, the uppermost of which is perforated. The unit
can be mounted on a boat which moves the revolving belt through
contaminated lake or seawater. As the belt moves through the
water, a differential pressure gradient is established across
the twin-layered belt which causes the oil-water system to flow
through the pores. The outer section preferably with less than
45 pores per linear inch traps the heavy more viscous oils,
while the lighter, less viscous oils are held in the second
foam layer with more than 60 pores per linear inch [88].

Oil is recovered, and of course the foam regenerated, by
a pair of counterclockwise moving squeeze rollers, Rl and R2,
operating at different face pressures on the loaded foams.
Sufficient light pressure is exerted first with roller Rl to
remove water, but not oil, which is collected in a trough and
returned to the main body of open water. Roller R2 then

squeezes the belts with greater force to remove adsorbed oil, which is collected and stored on the boat which moves over the water at speeds of 1/2 to 3 miles hr^{-1} [88].

3.8   ENZYME INHIBITORS AND IMMUNOADSORPTION OF CELLS

Open-cell polyurethane foams can be readily utilized to support starch gel containing enzymes, in particular, horse-serum cholinesterase [82-84]. This enzyme is first prepared as a starch gel and then immobilized on a polyurethane foam pad. Provided the cholinesterase remains active, the substrate, e.g., butyryl-thiocholine iodide, will be converted to the readily oxidizable product, thiocholine iodide:

$$(CH_3)_3N^+ \cdot I^- (CH_2)_2SCOC_3H_7 \xrightarrow{H_2O} (CH_3)_3N^+ \cdot I^- (CH_2)_2SH + C_3H_7COOH$$

(2)

The potential advantage of using immobilized enzyme in a foam pad located between two platinum electrodes (3 units per pad) over many hours was evaluated by passing substrate (5 × 10^{-4}M) in pH 7.4 tris buffer at 1 cm^3 min^{-1} and air at 1 dm^3 min^{-1}. Oxidation of the thiol to disulfide occurs at the platinum anode which was located downstream from the pad reactor where the hydrolysate concentration is greatest. At constant current (2 μA) a potential of about 150 mV was recorded across the cell while simultaneously passing substrate and air through the pad. If no enzyme is present in the pad due to inherent instability or ostensibly because of inactivation by air- or water-borne organophosphorus compounds, the potential will rise to that of the iodide-iodine couple, 350 to 400 mV. Thus, if the enzyme pad is completely active, a potential of about 150 mV will be recorded, but any loss of enzyme, by whatever agency, will produce a proportional rise in the potential until eventually, for total loss of enzyme, it reaches about 400 mV.

Unlike the classic assay technique with soluble enzyme,
the immobilized cholinesterase could be used continuously for
up to 12 hr to determine acetylcholine and butylthiocholine
based on calibration plots of log[substrate] vs. potential.
In addition the pad was used to detect anticholinesterase
inhibitors such as insecticides [83].

Enzyme pads incorporating aluminum hydroxide gel showed
higher enzyme activities and washout resistances. These improved pads have been used for monitoring organophosphorus
compounds in water supplies, and even after passing 2700 dm^3
of sample, useful amounts of enzyme remained on the foam pads
[82].

Coupling a specific antibody, e.g., antierythrocyte antibody, to a gum arabic coating solution on reticulated polyester
foam ($\sim$40 pores cm^{-1}) serves as a matrix for the immunological
binding of erythrocytes. The foam column is mounted horizontally
and slowly rotated about its long axis in order to permit maximum
contact between cells and foam structure while also allowing free
passage of unbound cells through the column. Thus, a single
antibody-coated foam (2.5 cm diameter and 1.25 cm thick) could
bind up to 3×10^8 guinea pig erythrocytes. The high specificity
of the method is further illustrated by the fact that erythrocytes coated with haptens were bound specifically to foam coated
with their corresponding antihapten antibodies with a low background of cross-reactivity [85].

4

Foam Chromatography
of Organic Compounds

Open-pore polyurethane foam has been utilized mostly in column
chromatography applications for gas-solid, gas-liquid, and low-
pressure liquid-liquid separations of organic compounds.

4.1 GAS-SOLID CHROMATOGRAPHY

Van Venrooy [1] first described the use of open-pore polyurethane for gas chromatography in 1967. The foam (13.2 g) with
about 560 pores per linear meter was cut into pieces <0.6 cm
and pushed into a copper tube (3 m long × 0.6 cm diameter).
Only moderate pressure was applied to achieve uniform packing
of the column so as to avoid crushing the cell structure and
thereby reduce the available surface area. This column attached
to a Beckman GC2A chromatograph and with helium as carrier gas
at 46 cm^3 min^{-1} facilitated the separation of the following
mixtures: water, acetic acid, and propionic acid in about 8 min;
and heptane, benzene, toluene, ethylbenzene, pentyl benzene, and
2-ethyl phenol in about 22 min [1].

Schnecko and Bieber ranked polyether and polyester polyurethane foams with foams of polyethene, styrene-butadiene
rubbers, and natural rubber as suitable gas-solid chromatography

fillers [11]. Thus, a mixture of nine C_6 to C_{14} aliphatic
hydrocarbons was easily separated with practically no tailing
on a polyether foam column (10 g foam, 2 m long × 0.6 cm
diameter) in about 30 min with nitrogen carrier gas and the
temperature programmed from 90 to 180°C. A very similar
pattern was obtained at 180°C with a mixture of water, methyl
and ethyl cellosolve, acetic acid, and ethylene glycol. These
examples emphasize that a nonpolar and polar mixture can be
resolved on one column. Both polyether and polyester foam
columns can also be employed for the analysis of the pyrolysis
products of various polymers [11]. In addition, gas chromatography was used to measure the glass transition temperature T_G
of the polymer used as the stationary phase in the column. The
value of -58°C for a polyether foam compared favorably with the
value obtained from the torsional modulus-temperature plot [11].

Other workers [3,5,6], unlike Van Venrooy [1] and Schnecko
and Bieber [11], prefer to employ chromatography columns composed
of in situ formed open-pore polyurethane. Typically [6] for this
purpose the chosen polyol-isocyanate ingredients dissolved in a
suitable dry solvent (toluene-carbon tetrachloride, 60:40 v/v)
was injected into the narrow glass column (0.4 cm i.d.). The
column with both ends clamped shut with Tygon tubing was rotated
vertically at 5 rpm for 18 hr at room temperature to complete
the polymerization process. Rotation prevents settling of the
polymer in a system where the polymer density does not match
the solvent density and also allows any bubbles of carbon dioxide
to rise to the top of the column [6]. In some cases the column
is closed only at the lower end. Solvent was then forced out
with dry nitrogen and the column placed in a chromatography
oven with its detector end disconnected and conditioned for
24 hr with gas flow at 100°C. Columns made of flexible copper
or aluminum may be similarly prepared in the linear configuration
and then bent to any desired shape before removing the solvent

with nitrogen gas. Columns with high sample capacities can
also be fabricated, e.g., 30 m long and 0.56 to 0.86 mm i.d.
and for preparative work may be up to 5.7 cm in diameter [6].

Electron photomicrographs show the polyurethane to be a
homogeneous conglomerate with microscopic spheres bonded
together and which also adhere to the walls of the glass column,
thereby preventing channeling at the wall-foam interface. All
the spheres measured in this rigid matrix were close to 4 μm in
diameter, and yet the permeability of the columns was comparable
to Chromosorb W columns [6]. In addition, the columns are
mechanically strong and with a thermal stability sufficient to
allow short-term temperature-programmed runs up to 200°C.

The density of the resultant foams is critical. Thus in
situ foams with densities <0.13 g cm^{-3} relate to poorly filled
columns with serious voids and channeling; for densities >0.18
g cm^{-3} columns have very low permeabilities and reduced plate
numbers. Foams with densities in the range 0.15 to 0.18 g cm^{-3}
are recommended for most work. Crowley has positioned in situ
formed polyurethane foams at intervals along chromatography
columns to reduce channeling [146]. However, it is far easier
to control channeling by the correct choice of polyol-isocyanate
ingredients prior to an in situ synthesis.

Such wall-to-wall in situ foam columns have permitted the
separation of a mixture of eight n-aliphatic hydrocarbons, e.g.,
C_6 to C_{10} and C_{12}, C_{14} and C_{16} [3,5]. One of the most interesting applications concerns the remarkable fast separation of
metal chelates [4,6]. Thus, the cis- and trans-isomers of
$Cr(TFA)_3$ where the ligand TFA represents 1,1,1-trifluoropentane-
2,4-dione are completely resolved on a 25-cm-long uncoated foam
column in <10 min (Fig. 1).

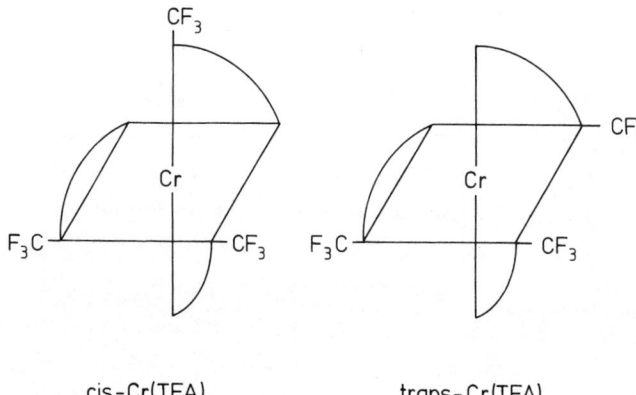

cis-Cr(TFA)$_3$ trans-Cr(TFA)$_3$

Kutal and Sievers have used this resolution to perform rate studies on the cis-trans equilibrium in the gas phase [8]. The corresponding geometrical isomers of rhodium have also been completely resolved in about 16 min [4].

That separations on these uncoated foams involve gas-solid adsorption was demonstrated by the peak shapes and retention

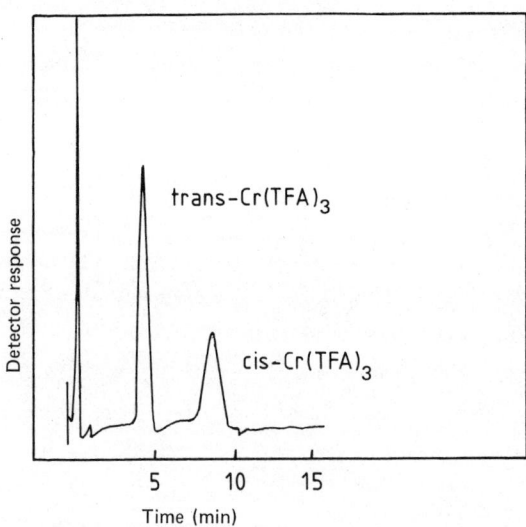

FIGURE 1   Gas chromatogram of cis- and trans-Cr(TFA)$_3$ isomers on uncoated polyurethane foam.   (From Ref. 4.)

times of different samples of dodecane [6]. Thus, the Van
Deemter plot showed no loss in plate number with flow rates as
high as 150 cm^3 min^{-1}. The adsorption process probably occurs
very rapidly on the uncoated foam surface which results in a
small mass transfer term C in the Van Deemter equation [6].

4.2 GAS-LIQUID CHROMATOGRAPHY

Stationary phases may be readily incorporated into foams either
by simply coating the actual foam or preferably by mixing with
the polyol-isocyanate reactants prior to injection into the
glass or metal column. This results in a more homogeneous
distribution of the stationary phase through the ultimate foam
product. Generally, DC-550 silicone oil, which showed a volatility loss of about 3% after 4 hr at 250°C, has been used for
this purpose [3]. Coating in this fashion masks the gas-solid
properties of the foam and the resultant material adopts the
character of a gas-liquid support with greater sample capacity
and efficiencies of 10 to 13 plates cm^{-1}. For optimum gas-
liquid chromatography performance the quantity of liquid phase
must exceed 5% m/m, but loadings > 50% m/m result in decreased
column efficiency because of the large C term in the Van Deemter
equation [6].

Further evidence for the masking of the highly polar character of the uncoated foam by the DC-550 silicone oil liquid
phase is shown in Fig. 2, where four metal chelates, $Be(TFA)_2$,
$Al(TFA)_3$, $Cr(TFA)_3$ and $Rh(TFA)_3$ are all clearly separated from
a mixture in about 5 min, whereas there is no resolution of the
cis- and trans-isomers [3,4,6]. It is also pertinent that the
primary aliphatic alcohols methanol, ethanol, propanol, butanol,
pentanol, and hexanol elute in order of boiling point from a
polyurethane foam coated with Carbowax 400 [6].

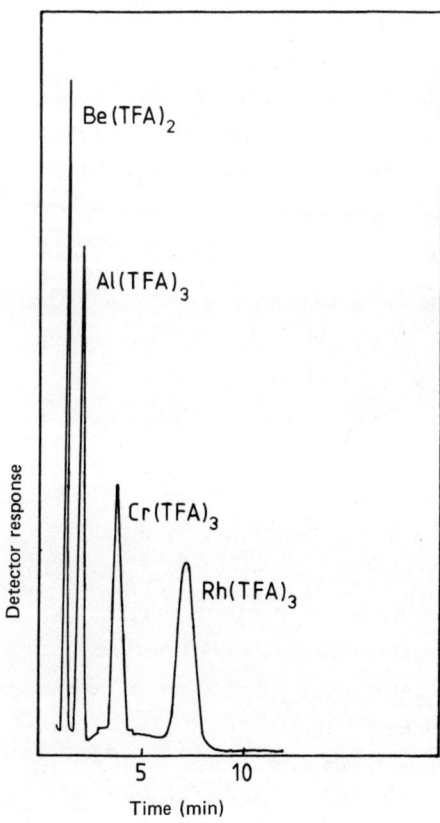

FIGURE 2   Gas chromatogram of four metal TFA chelates on polyurethane foam coated with 10% m/m DC-550.   (From Ref. 4.)

Other successful separations include mixtures of aliphatic
C_6 to C_9 hydrocarbons [3,4] and benzene, toluene, 1,2- and 1,3-
dimethyl benzenes [3].

4.3   LIQUID-LIQUID CHROMATOGRAPHY

Open-pore polyurethane foam is stable up to pressures of 105 kg
cm^{-2}, but its permeability is so good that it is unlikely to be
employed at such high pressures [4,10].   For example, columns

1 m long and 0.23 cm i.d. exhibit an inlet pressure of >7 kg
cm^{-2} at a flow rate of 1.0 cm^3 min^{-1} with most solvents [11].
High-speed analytical separations of benzene and phenol derivatives have been performed [7] at inlet pressures <0.21 kg cm^{-2}
(Table 1). Preparative-scale separations have also been achieved
on 50-mg samples of chlorophenols in less than 1 hr by using
1,2-dichloroethane-heptane (1:1 v/v) as a solvent [7]. The
need for expensive high-pressure pumps is therefore eliminated
in this approach to high-performance liquid chromatography.

In addition to favorable hydrodynamic considerations, the
mechanical properties of open-pore polyurethane particles (1 to
10 μm diameter) binding together facilitate the direct injection
of samples into the column material without blockage of syringe
needles [4]. Furthermore, the intimate wall-to-wall binding of
in situ foam columns avoids detector cell blockage and eliminates
the need for porous filters that contribute to band spreading
[10].

The capacities of certain open-pore foam preparations as
demonstrated by the 0.05-m column (Table 1) for separating a
mixture of benzene and chlorophenols indicate that columns need
only be lengthened in order to increase capacity. In this
example the phenol peak has a capacity factor of 3.4, and over
175 plates are generated by the 0.05-m column, that is, over
3500 plates per meter [7].

In contrast the alumina- and silica-type adsorbents used
for liquid chromatography, the open-pore polyurethane columns
have a wide dynamic loading range [10]. Thus, a column 1 m
long and 0.2 cm i.d. will take sample loads up to 5 μg of
2,4-dichloroaniline before peak broadening occurs. A further
increase in sample size causes no loss in peak symmetry. This
kind of behavior is more typical of a liquid-liquid chromatographic system than of a liquid-solid system [6].

TABLE 1 Liquid-Chromatographic Separations Using Open-Pore Polyurethane-Packed Columns

Column size (m length × cm i.d.)	OH-to-NCO ratio	Eluting solvent (proportions v/v)	Flow rate (cm^3 min^{-1})	Inlet pressure (kg cm^{-2})	Compounds separated in order of elution [time (min) in parenthesis]	Ref.
0.5 × 0.23	2.22	1,2-Dichloroethane	1.0	21	Benzene, 2-chlorophenol, phenol, and 4-chlorophenol (∿10)	[7][a]
0.05 × 0.23		Heptane + 1,2-dichloroethane (1:1)	0.4	0.21	Benzene, 2-chlorophenol, phenol, and 4-chlorophenol (∿5)	
0.5 × 0.23			0.85	4.2	Benzyl alcohol, cinnamyl alcohol, and 2-chlorophenol (∿10)	
0.5 × 0.23		Heptane + ethanol (70:30)	0.7	3.5	Benzene and nitrobenzene (∿5)	
1.0 × 0.2	1.5	Propan-2-ol + heptane (1:4)	0.26	4.9	Benzene, 2,6-dichloroaniline, 2,4-dichloroaniline, 2,3-dichloroaniline, 3,5-dichloroaniline, and 3,4-dichloroaniline (∿38)	[10][a]
				4.9	Benzene, acetone, and 2-aminophenol (∿35)	
	2.0			3.5	Benzene, 2,6-dichloroaniline, 2,4-dichloroaniline, 2,3-dichloroaniline, 3,5-dichloroaniline, and 3,4-dichloroaniline (∿33)	

[a]Detection with ultraviolet attachment.

Studies on solvent compatibility indicate that except for dimethylformamide a wide choice of solvents can be used with these columns [7,10]. However, toluene, chloroform, 1,2-dichloroethane, and tetrahydrofuran lead to mass losses of the polyurethane.

A mixture of the chelates of copper(II), cobalt(II), and cobalt(III) derived from 1,1,1,2,2,3,3-heptafluoro-7,7-dimethyl-4,6-octanedione (FOD) has been completely resolved on a 1-m-long polyurethane column [3,5]. Elution with benzene rapidly removed the dark green Co(FOD)$_3$ from the column. This was followed by the red band of Co(FOD)$_2$ on changing to ethanol, while the blue Cu(FOD)$_2$ complex remained at the top of the foam column.

The use of open-pore foam as a thin-layer chromatographic support has also been briefly reported for the separation of camptothecin from a crude extract of *Camptotheca acuminata* using benzene-acetone-methanol (16:4:0.5 v/v/v) as the developing solvent [3].

4.4 MODIFIED POLYURETHANE FOAMS

Foams for liquid and gas-solid chromatography have been prepared by either simply mixing with charcoal [72,73,75] or synthesized from isocyanate and cyclic oligosaccharides [147].

4.4.1 Charcoal-Dispersed Foams

Herbicide orange (HO) is primarily a mixture of n-butyl esters of 2,4-dichlorophenoxy acetic acid (2,4-D) and 2,4,5-trichlorophenoxy acetic acid (2,4,5-T). HO was used extensively as a military defoliant in Southeast Asia until banned following the discovery that 2,4,5-T contained 2,3,7,8-tetrachlorodibenzo-p-dioxin (TCDD)--one of the most toxic chemicals known.

TCDD

Since then other related polychlorinated dibenzo-p-dioxins
(PCDDs) have also been detected in 2,4,5-T and which are also
extremely toxic.

Plans are underway to dispose of stocks of herbicide orange
by incineration at sea, and trace analysis for such toxic compounds in HO samples is thus of environmental importance. While
PCDDs can be recovered using charcoal adsorption columns, exhaus-
tive extraction with an aromatic solvent is required. To overcome the flow restrictions associated with charcoal columns,
Huckins et al. [75] examined the prospective use of charcoal-
dispersed foams comprising 14.9% PX-21 charcoal + 85.1% foam m/m.
HO in chloroform was first percolated through the column (1.5 g
and 12 cm long x 1 cm i.d.) followed by benzene to elute material
which would otherwise interfere in the GC-ECD analysis of TCDD.
The TCDD and any related PCDDs were then eluted with 200 cm^3 of
toluene-benzene (1:1 v/v) and after careful concentration submitted for GC analysis. Control experiments with batches of HO
spiked with [^{14}C]-2,3,7,8-TCDD showed recoveries of 91.4% ± 4.0%
(n = 4).

Mirex, 1,1a,2,2a,3,3a,4,5,5a,5b,6-dodecachlorooctahydro-
1,3,4-metheno-2H-cyclobuta[cd]pentalene is a pesticide found in
Lake Ontario fish. Its toxicity, stability, and biological concentration has warranted monitoring for mirex and a suspected
degradation product, photomirex (8-monohydromirex). However,
PCBs which are frequently present interfere in the analysis for
mirex and photomirex by gas-liquid chromatography. To overcome
this problem new adsorbents based on the concept of fine mixtures
of various charcoals with polyurethane as developed for herbicide
analysis [75] were investigated [72,73]. No less than 14 different types of charcoal mixed with pesticide quality foam (Analab
Inc.) were examined for the adsorption-elution behavior of mirex,
photomirex, and Aroclor 1254. To offset the lengthy elution
times associated with charcoal products, the foam was shredded

[75] before mixing with the charcoal in chloroform.   After
removing the chloroform on a water bath, the foam-charcoal was
dried in a desiccator for 24 hr, sealed in jars, and stored in
a desiccator.   Glass columns (10 cm × 5 mm i.d.) were then
filled with 5 cm (0.2 g) of the foam-charcoal support (60:40%
m/m) and top-loaded with a standard mixture of mirex (1 µg),
photomirex (1 µg) and Aroclor 1254 (3 µg).   The first two compounds eluted with benzene free-cyclohexane (10 cm^3) and Aroclor
1254 with benzene (10 cm^3).   The high-quality cyclohexane is
important for this clear-cut separation since control runs with
cyclohexane spiked with varying amounts of benzene (500 to
50,000 ppm) affected the recoveries of the materials.   Thus,
the recovery of Aroclor 1254 in the second benzene fraction
declined as the benzene in the first eluting solvent (cyclohexane) was raised because considerable amounts of PCB were
eluted into this first fraction.   Therefore, traces of benzene
in cyclohexane are to be avoided for a clear-cut fractionation
of mirex and photomirex from PCBs [72].

      Thus, with proper grade cyclohexane both mirex and photo-
mirex GC peaks, but not Aroclor 1254, are clearly present in
the cyclohexane fraction submitted for gas-liquid chromatography.
Similarly only the PCB peaks show up in the benzene fraction
[72,73].

      The average recovery using Norit C-170 charcoal-based foam
was 100.1% (sd = 10.0) and 99.1% (sd = 9.9), respectively, for
mirex and photomirex at the 1-µg level and 96.8% (sd = 9.8) for
Aroclor 1254 at the 3-µg level [73].   For the other 13 charcoals
recoveries for similar duplicate assays were 71.9 to 109.7% and
47.0 to 115.5%, respectively, for mirex and photomirex and 23.3%
to 114.6% for Aroclor 1254.

4.4.2   Cyclodextrin-Polyurethane Foams

The gas-solid chromatography properties of polyurethane foams
prepared by condensing various isocyanates with α- and β-cyclo-

dextrins have been recently reported [147]. Cyclodextrins are
torus-shaped oligosaccharides comprising 1,4-linked glucopyranose
units. The Greek letters α and β represent the number of these
units, e.g., α is six and β is seven. The condensation shown
for one of these glucose ring units at C_6,

$$+ \text{RNCO} \tag{1}$$

is not complete as indicated by the OH residues per cyclodextrin-
foam molecule, e.g., 14.4 for the β-HDI-P-5.5-M product. In
this format for a typical preparation, β represents the β-cyclo-
dextrinpolyol; HDI, hexamethylene diisocyanate; 5.5 is the mass
of HDI(g) dissolved in solvent P (pyridine); and M (methanol)
the solvent employed to precipitate the final foam product from
the condensation reaction. The formats for other products are
covered in Table 2. If necessary the free primary and secondary
OH groups can be completely silanized with trimethylchlorosilane
in n-hexane at 60°C for 4 hr. The effect of silanization on the
retention times of five compounds is shown in Table 2 for the
β-HDI-P-5.5-M foam.

The retention times for a selection of compounds on four
different foams in Table 2 illustrates their considerable

TABLE 2 Retention Times (min) of Selected Compounds Relative to Benzene on Various Cyclodextrin-Type Foams at Two Different Temperatures

Compound	β-HDI-P-5.5-A[a] 150°C	β-HDI-DMF[b]5.5-A 150°C	β-HDI-DMF[b]5.5-A 170°C	β-HDI-P-5.5-M 170°C	β-HDI-P-5.5-M-Si[c] 170°C	α-HDI-DMF-5.9A 150°C	α-HDI-DMF-5.9A 170°C	α-HDI-DMF-13.3-A 150°C
Hexane	0.04	0.04				0.16		0.27
Heptane	0.09	0.06				0.24		0.33
Octane	0.15	0.10				0.29		0.45
Methanol	0.12	0.36				0.53		1.06
Ethanol	0.35	0.78		0.31	0.32	0.83		1.53
Propan-1-ol	1.33	1.96				2.03		2.80
Methylethyl ketone	1.03	1.24		0.81	0.60	1.61		1.65
Methyl propyl ketone		2.02				5.44		3.32
Methyl propanoate	0.69	0.85		0.54	0.37	1.84		1.20
Ethyl propanoate		1.11				6.80		1.90
Cyclohexane	0.05	0.05		0.13	0.15	0.07		0.12
Toluene		1.20	1.11			2.73	1.25	1.96
1,2-Dimethylbenzene		0.60	1.00			0.89	1.53	2.62
1,3-Dimethylbenzene		1.08	1.09			2.22	2.20	2.77
1,4-Dimethylbenzene		1.87	1.49			6.53	4.93	3.41
Benzene	1.00 (60.52)[d]	1.00 (16.68)	1.00 (7.25)	1.00 (28.76)	1.00 (22.32)	1.00 (6.59)	1.00 (3.88)	1.00 (4.16)

[a]Acetone.
[b]Dimethylformamide.
[c]Silanized product.
[d]Actual retention times in parentheses.
Source: Ref. 147.

127

prospect for the separation of many different mixtures [147].
The isomeric picolines and lutidines could be resolved using
these modified foam columns.

References

1. J. J. Venrooy, U.S. Patent 3,347,020, 1967.
2. R. T. Jefferson and I. O. Salyer, U.S. Patent 3,574,150, 1971.
3. I. O. Salyer, R. T. Jefferson, and W. D. Ross, U.S. Patent 3,580,843, 1971.
4. W. D. Ross, Open Pore Polyurethane--a New Separation Medium, in *Separation and Purification Methods*, Volume III, No. 1, Dekker, New York, 1974, p. 111.
5. W. D. Ross and R. T. Jefferson, *J. Chromatog. Science*, 8: 386 (1970).
6. F. D. Hileman, R. E. Sievers, G. G. Hess, and W. D. Ross, *Anal. Chem.*, 45: 1126 (1973).
7. L. C. Hansen and R. E. Sievers, *J. Chromatog.*, 99: 123 (1974).
8. K. C. Kutal and R. E. Sievers, *166th National A.C.S. Meeting, Chicago*, American Chemical Society, Washington, D.C., 1973.
9. J. D. Navratil and R. E. Sievers, *International Laboratory*, Nov.-Dec.: 26 (1977).
10. T. R. Lynn, D. R. Rushneck, and A. R. Cooper, *J. Chromatog. Science*, 12: 76 (1974).
11. H. Schnecko and O. Bieber, *Chromatographia*, 4: 109 (1971).
12. H. J. M. Bowen, *J. Chem. Soc.*, (A): 1082 (1970).
13. H. J. M. Bowen, British Patent, 1,305,375, 1973.
14. I. Valente and H. J. M. Bowen, *Analyst*, 102: 842 (1977).
15. I. Valente and H. J. M. Bowen, *Anal. Chim. Acta*, 90: 315 (1977).
16. S. Sukiman, *Radiochem. Radioanal. Lett.*, 18: 129 (1974).
17. T. Braun and A. B. Farag, *Talanta*, 19: 828 (1972).
18. T. Braun and A. B. Farag, *Talanta*, 22: 699 (1975).

19. T. Braun and A. B. Farag, *Anal. Chim. Acta*, 68: 119 (1974).
20. T. Braun and A. B. Farag, *Anal. Chim. Acta*, 66: 419 (1973).
21. T. Braun and A. B. Farag, *Anal. Chim. Acta*, 61: 265 (1972).
22. T. Braun, L. Bakos, and Z. S. Szabó, *Anal. Chim. Acta*, 66: 57 (1973).
23. T. Braun, A. B. Farag, and A. Klimes-Szmik, *Anal. Chim. Acta*, 64: 71 (1973).
24. T. Braun, É. Huszár, and L. Bakos, *Anal. Chim. Acta*, 64: 77 (1973).
25. T. Braun and A. B. Farag, *Anal. Chim. Acta*, 65: 115 (1973).
26. T. Braun and A. B. Farag, *Anal. Chim. Acta*, 73: 301 (1974).
27. T. Braun and A. B. Farag, *Anal. Chim. Acta*, 69: 85 (1974).
28. T. Braun and A. B. Farag, *Anal. Chim. Acta*, 71: 133 (1974).
29. T. Braun and A. B. Farag, *Anal. Chim. Acta*, 76: 107 (1975).
30. T. Braun and A. B. Farag, *Radiochem. Radioanal. Lett.*, 19: 275 (1974).
31. T. Braun and A. B. Farag, *Radiochem. Radioanal. Lett.*, 19: 377 (1974).
32. T. Braun and A. B. Farag, *J. Radioanal. Chem.*, 25: 5 (1975).
33. T. Braun and A. B. Farag, *Anal. Chim. Acta*, 65: 139 (1973).
34. S. Palágyi and T. Braun, *J. Radioanal. Chem.*, 51: 267 (1979).
35. T. Braun and A. B. Farag, *Anal. Chim. Acta*, 98: 133 (1978).
36. T. Braun, A. B. Farag, and M. P. Maloney, *Anal. Chim. Acta*, 93: 191 (1977).
37. T. Braun and M. N. Abbas, *Anal. Chim. Acta*, 119: 113 (1980).
38. M. P. Maloney, G. J. Moody, and J. D. R. Thomas, *Proc. Anal. Div. Chem. Soc.*, 14: 244 (1977).
39. M. P. Maloney, G. J. Moody, and J. D. R. Thomas, *Analyst*, 105: 1087 (1980).
40. M. P. Maloney, Ph.D. Thesis, Wales, 1980.
41. T. Braun and S. Palágyi, *Anal. Chem.*, 51: 1697 (1979).
42. T. Braun and M. N. Abbas, *Anal. Chim. Acta*, 134: 321 (1982).
43. S. Palágyi and E. Bilá, *Radiochem. Radioanal. Lett.*, 32: 87 (1978).
44. S. Palágyi and R. Markusová, *Radiochem. Radioanal. Lett.*, 32: 103 (1978).
45. S. Palágyi and T. Braun, *180th National A.C.S. Meeting, Las Vegas*, American Chemical Society, Washington, D.C., 1980.
46. R. A. Moore and A. Chow, *Talanta*, 27: 315 (1980).
47. H. D. Gesser, E. Bock, W. G. Baldwin, A. Chow, D. W. M. McBride, and W. Lipinsky, *Separation Science*, 11: 317 (1976).
48. H. D. Gesser and G. A. Horsfall, *J. Chim. Phys.*, 74: 1072 (1977).
49. P. R. Musty and G. Nickless, *Proc. Anal. Div. Chem. Soc.*, 12: 295 (1975).
50. H. J. M. Bowen, *Radiochem. Radioanal. Lett.*, 7: 71 (1971).

51. K. Srikameswaran and H. D. Gesser, *J. Environ. Sci. Health,*
     A13: 415 (1978).
52. D. W. Lee and M. Halmann, *Anal. Chem.,* 48: 2214 (1976).
53. P. Schiller and G. B. Cook, *Anal. Chim. Acta,* 54: 364
     (1971).
54. G. N. Lypka, H. D. Gesser, and A. Chow, *Anal. Chim. Acta,*
     78: 367 (1975).
55. A. Chow and D. Buksak, *Can. J. Chem.,* 53: 1373 (1975).
56. M. A. J. Mazurski, A. Chow and H. D. Gesser, *Anal. Chim.
     Acta,* 65: 99 (1973).
57. V. S. K. Lo and A. Chow, *Talanta,* 28: 157 (1981).
58. A. Chow, G. T. Yamashita, and R. F. Hamon, *Talanta,* 28:
     440 (1981).
59. G. J. Moody and J. D. R. Thomas, *Analyst,* 104: 1 (1979).
60. G. J. Moody, J. D. R. Thomas, and M. A. Yarmo, unpublished
     work.
61. P. R. Musty and G. Nickless, *J. Chromatog.,* 120: 369 (1976).
62. P. R. Musty and G. Nickless, *J. Chromatog.,* 100: 83 (1974).
63. H. D. Gesser, A. Chow, F. C. Davis, J. F. Uthe, and J.
     Reinke, *Anal. Lett.,* 4: 883 (1971).
64. H. D. Gesser, A. B. Sparling, A. Chow, and C. W. Turner,
     J. Amer. Water Works Ass., 65: 220 (1973).
65. K. M. Gough and H. D. Gesser, *J. Chromatog.,* 115: 383 (1975).
66. J. F. Uthe, J. Reinke, and H. D. Gesser, *Environ. Lett.,* 3:
     117 (1972).
67. H. D. Gesser, A. Chow, F. C. Davis, J. F. Uthe, and J.
     Reinke, *P.C.B. Newsletter,* 4: 11 (1972).
68. J. F. Uthe, J. Reinke, and H. O'Brodovich, *Environ. Lett.,*
     6: 103 (1974).
69. C. G. Simon and T. F. Bidleman, *Anal. Chem.,* 51: 1110 (1979).
70. T. F. Bidleman and C. E. Olney, *Bull. Environ. Contam.
     Toxicol.,* 11: 442 (1974).
71. T. F. Bidleman and C. E. Olney, *Science,* 183: 516 (1974).
72. A. J. Y. Chau and L. J. Babjack, *J. Ass. Off. Anal. Chem.,*
     62: 107 (1979).
73. L. J. Babjack and A. S. Y. Chau, *J. Ass. Off. Anal. Chem.,*
     62: 1174 (1979).
74. J. W. Bedford, *Bull. Environ. Contam. Toxicol.,* 12: 622
     (1974).
75. J. N. Huckins, D. L. Stalling, and A. Smith, *J. Ass. Off.
     Anal. Chem.,* 61: 32 (1978).
76. C. L. Stratton, S. A. Whitlock, and J. M. Allan, *U.S.
     Environmental Protection Agency Research Report,* EPA-60014-
     78-048, 1978.
77. T. J. Murphy and C. P. Rzeszutko, *J. Great Lakes Res.,* 3:
     305 (1977).
78. H. Yamasaki and K. Kuwata, *Bunseki Kagaku,* 26: 1 (1977).
79. B. C. Turner and D. E. Glotfelty, *Anal. Chem.,* 49: 7 (1977).

80. R. G. Lewis, A. R. Brown, and M. D. Jackson, *Anal. Chem.*,
     49: 1668 (1977).
81. P. V. Doskey and A. W. Andrew, *Anal. Chim. Acta*, 110: 129
     (1979).
82. L. H. Goodson, W. B. Jacobs, and A. W. Davis, *Anal. Bio-
     chem.*, 51: 362 (1973).
83. E. K. Bauman, L. H. Goodson, G. G. Guilbault, and D. N.
     Kramer, *Anal. Chem.*, 37: 1378 (1965).
84. E. K. Bauman, L. H. Goodson, and J. R. Thomson, *Anal.
     Biochem.*, 19: 587 (1967).
85. W. H. Evans, M. G. Mage, and E. A. Peterson, *J. Immunol.*,
     102: 899 (1969).
86. R. C. Schlicht and F. C. McCoy, U.S. Patent, 3,617,531,
     1971.
87. R. L. Strickman, U.S. Patent, 3,618,618, 1971.
88. R. G. Will and J. F. Grutsch, U.S. Patent, 3,617,552, 1971.
89. E. C. A. Cadron and L. E. C. Jourquoin, Belg. Patent,
     764,805, 1971.
90. C. R. Thomas, *British Plastics*, 552 (1965).
91. *Proc. Conf. on Cellular Plastics, Mass., 1966*, NAS/NRC,
     Washington, D.C., 1967.
92. T. Braun, O. Békeffy, I. Haklits, K. Kádár, and G. Majoros,
     Anal. Chim. Acta, 64: 45 (1973).
93. F. D. Hileman, M.S. Thesis, Wright State University, 1973.
94. S. Gross (ed.), *Modern Plastics Encyclopaedia*, Vol. 46,
     McGraw-Hill, New York, 1969.
95. T. H. Ferringo, *Rigid Plastic Foams*, Reinhold, New York,
     1963.
96. P. J. Corish, *Anal. Chem.*, 31: 1298 (1959).
97. E. B. Murphy and W. A. O'Neil, *S. P. E. J.*, 18: 191 (1962).
98. G. G. Greth, R. G. Smith, and G. O. Rudkin, *J. Cell Plas-
     tics*, 1: 159 (1965).
99. R. Merton, G. Braun, and D. Lauerer, *Kunstoffe*, 55: 249
     (1965).
100. L. I. Kuposov and V. V. Zharkov, *Plast. Massy*, 9: 63 (1972).
101. M. Sumi, Y. Chokki, Y. Naki, M. Nakabayashi, and T. Kanzawa,
     Makromol. Chem., 78: 146 (1964).
102. E. G. Brame, R. C. Ferguson, and G. J. Thomas, *Anal. Chem.*,
     39: 517 (1967).
103. G. K. Sankha and Y. Alaire, *Toxicol. App. Pharmacol.*, 50:
     533 (1979).
104. G. K. Sankha, M. Matijak, and Y. Alaire, *Toxicol. App.
     Pharmacol.*, 57: 241 (1981).
105. C. J. Purnell and R. F. Walker, *Anal. Proc.*, 18: 472 (1981).
106. J. Moreton and N. A. R. Falla, *Analysis of Airborne Pollu-
     tants in Working Atmospheres. The Welding and Surface
     Coating Industries*, The Chemical Society, London, 1980.
107. J. L. Guthrie and R. W. McKinney, *Anal. Chem.*, 49: 1676
     (1977).

108. D. J. H. Edwards, *J. High Resolut. Chromatog. Chromatog. Commun.*, 3: 190 (1980).
109. D. Lai, J. R. Arnold, and B. L. K. Somayajulu, *Geochim. Cosmochim. Acta*, 28: 1111 (1964).
110. H. J. M. Bowen, *Radiochem. Radioanal. Lett.*, 2: 169 (1969).
111. V. S. K. Lo and A. Chow, *Anal. Chim. Acta*, 106: 161 (1979).
112. T. Tanaka, K. Hiiro, and A. Kawahara, *Bunseki Kagaku*, 22: 523 (1973).
113. A. Baghai and H. J. M. Bowen, *Analyst*, 101: 661 (1976).
114. H. D. Gesser, G. A. Horsfall, K. M. Gough, and B. Krawchuk, *Nature*, 268: 322 (1977).
115. J. J. Oren, K. M. Gough, and H. D. Gesser, *Can. J. Chem.*, 57: 2032 (1979).
116. R. F. Hamon, A. S. Khan, and A. Chow, *Talanta*, 29: 313 (1982).
117. R. F. Hamon and A. Chow, *Abstracts of 60th C.I.C. Conference*, Chem. Institute of Canada, Montreal, 1977.
118. T. Braun and A. B. Farag, *Anal. Chim. Acta*, 99: 1 (1978).
119. R. F. Hamon, Ph.D. Thesis, Manitoba University, 1981.
120. G. H. Morrison and H. Freiser, *Solvent Extraction in Analytical Chemistry*, Wiley, New York, 1957.
121. C. J. Pedersen, *Federation Proc.*, 27: 1305 (1968).
122. C. J. Pedersen, *J. Amer. Chem. Soc.*, 89: 7017 (1967).
123. I. Sotobayashi, S. Tonouchi, and T. Suzuki, *Chem. Lett.* (Chem. Soc. Jap.), 585 (1976).
124. P. G. Delduca, A. M. Y. Jaber, G. J. Moody, and J. D. R. Thomas, *J. Inorg. Nucl. Chem.*, 40: 187 (1978).
125. R. Iwamoto, Y. Saito, H. Ishihara, and H. Tadokoro, *J. Polymer Sci.*, A2, 6: 1509 (1968).
126. L. Tusek, H. Meider-Gorican, and P. R. Danesi, *Z. Natürforsch.*, 318: 330 (1976).
127. H. Tadokoro, *Macromolecular Revs.*, 1: 119 (1967).
128. J. J. Christensen, D. J. Eatough, and R. M. Izatt, *Chem. Revs.*, 74: 351 (1974).
129. A. M. Y. Jaber, G. J. Moody, and J. D. R. Thomas, *J. Inorg. Nucl. Chem.*, 39: 1689 (1977).
130. K. Srikameswaran, H. D. Gesser, and M. Venkakateswaran, *J. Environ. Sci. Health*, A15: 323 (1980).
131. F. Feigl and V. Anger, *Spot Tests in Inorganic Analyses*, 6th ed., Elsevier, Amsterdam, 1972.
132. Y. K. Lee, K. J. Whang, and K. Ueno, *Talanta*, 22: 535 (1975).
133. Y. K. Lee, K. J. Whang, and K. Ueno, *Talanta*, 23: 244 (1976).
134. W. J. McDowell and C. F. Coleman, *J. Inorg. Nucl. Chem.*, 28: 1083 (1966).
135. A. Craggs, G. J. Moody, and J. D. R. Thomas, *Analyst*, 104: 412 (1979).
136. D. C. Gregoire and A. Chow, *Talanta*, 22: 453 (1975).
137. Anonymous, *Chem. Eng. News*, 49: 15 (1971).

138. O. Hutzinger, S. Safe, and V. Zitko, *The Chemistry of PCB's*, CRS Press, Cleveland, 1974.

139. J. W. Bedford, E. W. Roelofs, and M. J. Zabik, *Limnol. Oceanogr.*, 13: 118 (1968).

140. P. A. Butler, *Amer. Fish. Soc.* Special Publication No. 3, 1966.

141. G. E. Smith and B. G. Isom, *Pest. Monit. J.*, 1: 16 (1967).

142. B. Ahling and S. Jensen, *Anal. Chem.*, 42: 1483 (1970).

143. J. D. Millar, R. E. Thomas, and H. J. Schattenberg, *Anal. Chem.*, 53: 214 (1981).

144. R. G. Lewis and N. J. Zimmerman, *Anal. Qual. Contr. Newslett.*, U.S. EPA, 28: 7 (1976).

145. J. S. Armour, *J. Ass. Off. Anal. Chem.*, 56: 987 (1973).

146. R. P. Crowley, U.S. Patent, 3,422,605, 1969.

147. Y. Mizobuchi, M. Tanaka, and T. Shono, *J. Chromatog.*, 194: 153 (1980).

Index

Acetylcholine, 113
Active carbon, 58
Alamine-336, 70
Alcohols in foam chromatog-
    raphy, 119, 122, 126
Aldrin, 81, 88-90
Alkylbenzene sulfonate, 106
Allophanate links, 12, 13
Al(TFA)$_3$, 119
Amberlite:
  LA-1, 67
  LA-2, 68
Anchored foams, 74
Aroclor:
  1016, 79, 81, 82, 93-98, 100
  1221, 81, 82, 96, 98
  1232, 81
  1242, 79-82, 96, 98-101
  1248, 79, 81, 98, 101
  1254, 79, 81, 82, 96, 98,
    100, 101, 123
  1260, 81, 86, 87, 98, 100
Atlantic PCBs, 100
Atomic absorption spectroscopy,
    76

Be(TFA)$_2$, 119
Benzo(a)pyrene, 104
Benzoyl acetone, 64

α-BHC, 81, 87-89
β-BHC, 81, 88, 89
γ-BHC, 81, 82, 87-90
δ-BHC, 81
Biphenyl, 98
Biuret links, 12, 13
Blowing agents, 8, 110
Breakthrough capacity, 53, 121
Bromononachlorobiphenyl, 98
Butyrylthiocholine, 113

Camptothecin, 123
Capacity, 20, 74, 121
Carbamic acid, 3
Carbowax 400, 1, 119
Catalysts, 8, 110
Cation chelation mechanism, 45
Channeling, 117
Charcoal adsorption, 83
  dispersed foam, 123
Chelate foams, 60-68
Chloranil on foams, 74
Chlordane, 80, 81, 101, 104
Chloride media, 34, 56
Chlorobiphenyls, 79-102
p-chloronitrobenzene, 103
Chlorophenols by foam chromatog-
    raphy, 121
Cholinesterase, 113

Chromatography, 115-128
  gas-solid, 115
  gas-liquid, 119
  liquid-liquid, 120
  reversed phase foam, 51, 57
Chromofoams, 67
Chromosorb W, 117
Cigarette smoke, 110
Cis-Cr(TFA)$_3$, 117, 119
Cis-Rh(TFA)$_3$, 117
Classification, 4
Co(FOD)$_2$, 123
Co(FOD)$_3$, 123
Collection unit:
  PCBs from air, 91
  oil from water, 112
Cresols, 108
Crown ethers, 42
Crystal violet, 106
Cu(FOD)$_2$, 123
Cyclodextrin-polyurethane
    foams. 125

Dacthal, 102
4,4'-DDD, 81, 88, 89
4,4'-DDE, 81, 88-90
DDT, 80
2,4'-DDT, 99
4,4'-DDT, 81, 83, 88-90, 99, 101
Derivatization, 85
2,4-diaminotoluene, 16
2,6-diaminotoluene, 16
Diazinon, 94
Dieldrin, 81, 88-90
Diethyldithiocarbamate, 63
Di(2-ethylhexyl)phosphoric
    acid, 75
Dimethylglyoxime, 49, 64, 78
Disparlure, 104
Distribution ratio, 20
Dithizone, 60, 67, 72
Dodecane, 117

Efficiency, 72
Electronic spectra:
  of Co(NCS)$_4^-$-foam, 41, 45
  of FeCl$_4^-$-foam, 41

[Electronic spectra]
  of iodine-foam, 23
  of Pd(SCN)$_4^{2-}$-foam, 41
Endosulfan I and II, 81
Endosulfate, 81
Endrin 81, 88-90
Endrinaldehyde, 81
Enzyme inhibitors, 113
Erythrocytes, 113
Esters in foam chromatography,
    126
1,2-ethanedithiol, 63
Extractions on unloaded foams,
    19-49, 86
  antimony, 21, 27, 29, 43
  benzene, 21
  cadmium, 27, 32
  chloroform, 21
  cobalt, 27, 33, 35, 37, 40-
    43, 47, 49
  gallium, 27, 32, 38, 41, 43
  gold, 21, 25, 27
  group II metals, 27
  indium, 27, 35
  iodine, 21, 72
  iridium, 25, 27, 39
  iron, 21, 27, 32, 35, 37, 40,
    43
  lead, 32
  manganese, 32
  mercury, 21, 27, 32, 35
  molybdenum, 21
  nickel, 27, 32
  palladium, 43
  pesticides, 86
  phenol, 21
  platinum, 27, 29
  potassium, 49
  rhenium, 21
  rhodium, 25
  silver, 25, 49
  styrene, 21
  thallium, 21, 49
  tin, 27, 31, 37, 43
  uranium, 21, 39
  zinc, 27, 35
Extractions/separations on
    modified foams, 49-78, 86
  antimony, 52, 61, 63

[Extractions/separations on
    modified foams]
  bismuth, 52, 54, 56
  cadmium, 61, 63, 64, 66, 67,
    69, 73
  calcium, 73, 75
  cerium, 75
  cobalt, 52, 56, 65, 67, 69
  copper, 56, 61, 64, 69, 73
  gold, 52, 57, 59
  indium, 65
  iodine, 69
  iron, 52, 56, 58, 65, 68, 69,
    73, 75
  lead, 63, 67
  manganese, 65
  mercury, 50, 60, 61, 65, 67,
    70, 72, 74
  methyl mercury, 60
  nickel, 49, 52, 54, 56, 61,
    64, 69
  palladium, 51, 52, 54, 78
  pesticides, 86
  rhodium, 78
  silver, 52, 61, 63, 74
  tin, 67, 70, 78
  vanadium, 75
  zinc, 52, 58, 61, 65, 67, 69,
    73

Field tests for PCBs, 99
Flow rates, 50, 74, 102, 105,
    121
Functionality, 6, 7, 74

Glass transition temperatures,
    45, 116

Halowax 1001, 92
Health hazards, 14
Heptachlor, 81, 89, 90, 104
  epoxide, 81, 90
1,1,1,2,2,3,3-heptafluoro-7,7-
    dimethyl-4,6-octanedione
    (FOD), 123

Herbicide orange, 123
Herbicides, 102
Heterogeneous ion-exchange
    foams, 73
HETPs, 53, 119, 121
Hofmeister series, 49
Hydrocarbons in foam chroma-
    tography, 116, 117, 119,
    121, 122, 126

Immunoadsorption of cells, 113
Index format, 11
Infrared spectroscopy, 12, 17,
    44, 46, 50
Insecticides, 79
Ion-exchanger foams, 68, 73, 74
Ion-selective electrode monitor-
    ing, 76
Ionorganic compounds on foams,
    19-78
Isocyanates:
  4,4'-diphenylisocyante, 4,4'-
    diphenylmethane diisocyanate,
    2,4-toluenediisocyante, 2,6-
    toluenediisoyanate, 1,5-
    naphthalene diisocyanate, 7
  hexamethylene diisocyanate,
    7, 126
  health hazards, 14

Ketones in foam chromatography,
    126

Lake Michigan, 80
Land fill analysis for PCBs, 99
Lauryl benzene sulfonate, 106
Leaching effects, 65, 76
Ligand exchange, 41
Lindane, 81, 82, 87-90
Liquid exchangers, 67, 68
Loaded foams, 49-78
Lutidenes, 128

Malathion, 94
Mechanistic aspects, 36, 50

Metal-thiocyanate systems, 32
Methylene blue sorption, 14,
    86, 88, 107
Milk, iodine extraction, 70
Mirex, 124
2-monochlorobiphenyl, 96, 98
Mussels, 83

Natural sponge, 19
NCO/OH ratios, 8, 9
Nicotine, 110
1-nitroso-2-naphthol, 64
Nmr spectroscopy, 12, 13

Oils, 111
Open-pore polyurethane, 8, 115
Organic compounds on foam, 79-
    114
Oysters, 83

PAN, 64
Pattern matching, 97
Perchlorination, 97, 99
Pesticides, 79
Phenols, 107
Phoromone, 104
Photomirex, 124
Phthalate esters, 104
Picolines, 128
Polychlorobiphenyl (PCBs), 79-
    102
  decachlorobiphenyl, 97, 99
  2,2'-, 2,4-, 2,4'- and
    2,2',5-PCBs, 98
  3,3'-PCB, 93, 94
  2,4,5'-PCB, 93
  2,5,4'-, 3,4,3',4'- and
    2,4,6,2'4'6'-PCBs, 94
  3,4,2'-PCB, 92, 94
  2,5,2',5'-PCB, 92, 94-97
  2,4,5,2',5'-PCB, 92, 94
Polychloronaphthalenes, 92, 96
Polythene oxides, 42
Polyols, 6, 12, 13, 110
Polypropene oxides, 43

Polyurethanes:
  chemical properties, 13
  closed cell (rigid) type, 4
  compression strength, 10, 120
  density, 10, 117
  open cell (flexible) type, 4
  particle size, 5, 10
  porosity, 10, 117, 121
  1-(2-pyridylazo)-2-naphthol,
    64
  shore A hardness, 10
  surface area, 10, 19, 20
  synthesis, 2, 116
Precipitates in foams, 73
Properties of foams, 8-17, 117,
    120, 121
Pulsating columns, 70

Redox foams, 74
Reverse phase foam chromatog-
    raphy, 51, 57
Rh(TFA)$_3$, 119

Sargasso Sea, 83, 100
Silicone oils:
  DC-11, 89
  DC-200, 86, 89, 95
  DC-550, 119
  QF-1, 89
  SE-30, 89, 95
Silicone rubber foams, 78
Solvent extractants, 51, 64,
    83, 86
  loaded foams, 51, 59
Sponges, 19
Stability of foams, 13, 78
Structure of foams, 5
Synthesis of foams, 2, 6, 116

2,3,7,8-tetrachlorodibenzo-p-
    dioxin, 123
Thicholine, 113
Thiocyanato systems, 113
Toxaphene, 81
Trans-Cr(TFA)$_3$, 117, 119

Trans-Rh(TFA)$_3$, 117
1,1,1-trifluoropentane-2,4-
    dione (TFA), 117
Trifluralin, 102, 104
Tri-n-butyl phosphate, 50-52,
    54, 56, 58, 63
Tri-n-octylamine, 78

Unloaded foams, 19-49, 86
Urea links, 12, 13
Urethane links, 12, 13

Van Deemter plot, 119

X-ray diffraction, 44